THE TOOLS OF SCIENCE

THE TOOLS OF SCIENCE

Ideas and Activities for Guiding Young Scientists

Jean Stangl

Dodd, Mead & Company • New York

Published by Dodd, Mead & Company, Inc.
71 Fifth Avenue, New York, NY 10003
Distributed in Canada by
McClelland and Stewart Limited, Toronto
Manufactured in the United States of America
Designed by Suzanne Babeuf
First Edition

1 2 3 4 5 6 7 8 9 10

Library of Congress Cataloging-in-Publication Data

Stangl, Jean.
The tools of science.

Includes index.
1. Science—Study and teaching (Elementary)—United
States—Handbooks, manuals, etc. I. Title.
LB1585.3.S72 1987 372.3'5044 86-29197
ISBN 0-396-08965-8

ISBN 0-396-08966-6 {PBK.}

This book is dedicated to
Anna-Lee
Kenna
Joe
My young scientists

CONTENTS

PREFACE

Providing science experiences that will maintain a high interest level, stimulate curiosity, and motivate children to explore, experiment, and discover—these are fascinating and challenging tasks for adults today. This book shows you how you can introduce children to the wonderful world of science by providing them with opportunities to discover the importance of basic science tools, how to use them, and ways to make their own.

All the activities in this book provide hands-on experiences, and many are open ended. The experiments are fun to do, produce interesting results, and allow for recording predictions and observations. Equipment and materials consist mainly of recyclable and inexpensive, easy-to-obtain items.

Although designed primarily for classroom teachers and aides or student teachers (with or without formal science training) who are working with children in the third through the sixth grade, *The Tools of Science: Ideas and Activities for Guiding Young Scientists* will also be a valuable resource guide for any adult working with younger children, gifted students, older problem learners, and kids' clubs and organizations.

The variety of interesting activities for experimenting with science instruments, the easy-to-follow directions for making workable tools, and the simple explanations and think questions will all have special appeal for your young scientists.

The reproducible "Think and Do" sheets will encourage your beginning scientists to keep records of their experiments. You can use these valuable worksheets as a basis for a youngster's own lab book. The stimulating "Science Is Fun" section, containing quizzes, puzzles, and fascinating facts, will provide additional information and reinforce concepts in science.

A short review of children's science magazines and notations on those accepting contributions from children make up the appendix—it can lead to happy hours of reading for kids of all ages.

With a primary focus on the tools of science, these projects and activities will also encourage young scientists to observe, experiment, and solve problems as they discover the wonders of science in today's world.

To Steve Stangl and Barbara Mudrinich for their invaluable suggestions and editing assistance, I wish to extend my heartfelt thanks.

I also wish to thank Dr. Ruth Smith, who, unbeknownst to her, helped to make this book a reality.

And last, but by no means least, I wish to thank Mary Kennan for believing in me and my idea, for her patience and understanding and for giving me a second chance. This is her book too.

PART ONE

CREATING AN ENVIRONMENT FOR SCIENCE FUN . . . AND LEARNING

CHAPTER 1

A CENTER FOR LEARNING ABOUT SCIENCE

SETTING UP A CENTER

A place to discover, explore, experiment, and find out—these are the primary attributes of an effective science center, but almost anything can go into the center. Think of it not as a learning center but as a center of interest—or, better yet, a discovery center. It can be an entire room or one special corner. If your center holds children's interest and allows for discovery, they will learn.

Plan your center to focus on a single aspect of science and to provide an interesting variety of materials that are stimulating and motivating. After a few days, add additional tools, objects, or materials. Provide living and nonliving materials and introduce as many disciplines as possible, including biology, astronomy, and chemistry. When interest begins to wane, remove all materials and set up a new center. Be flexible enough to capitalize on the weather, the seasons, a sudden nature discovery, even objects brought from home. Science "work" should not be considered simply science, but rather experiences that will enable children to react with their environment in such a way as to gain practice in thinking, predicting, relating, observing, expressing, and cooperating.

When you don't know the answer to a child's question, admit it. And then help the inquirer to find the answer. Do your own research and share your new information as soon as possible. Never let a good learning situation slip by.

Make the science area interesting and attractive. Although science should not be an isolated subject, an area within the classroom should be set aside where science experiments can be carried out without interruption.

Stain-proof tables (or tables covered with a heavy sheet of plastic), a small tub for washing utensils, and sponges for cleanup should be a permanent part of the science area.

Establish and enforce rules and guidelines from the beginning. For maximum benefit to the children, allow them as much freedom as possible. A short explanation, the brief introduction of a new tool, or perhaps a simple reminder is all that should be necessary. If the experiment requires a lengthy demonstration or too much adult assistance, then the children are not ready for it, and you should choose an easier project.

You don't have to be a scientist or a science teacher to help children make discoveries and learn science. Neither is an understanding of all the scientific principles or expensive equipment necessary.

Adults should always be available to answer questions and provide resource information, but they should not hover over the youngsters or in any way stifle their interest in exploring and creating. Restrain yourself from "showing and telling" them how to do it.

Children have a natural sense of curiosity; they want to find out for themselves. This book is designed to provide a means for fostering that curiosity by allowing the child to become actively involved in both using and making tools. About 90 percent of learning how to think, make decisions, and do things is by actually doing. Children will thus learn best with hands-on activities. According to Piaget, "Each time we try to teach a child something that he could have discovered for himself, we are keeping the child from inventing it and consequently from understanding it completely."

We don't always need a predetermined goal, but all activities should have a purpose and be of value to the child.

Research has shown that the use of rewards and punishments in a learning situation tends to produce undesirable behavior and dislike for the subject matter, and it reduces the potential for learning. Set up the science center to enable your learners to be successful. If children feel successful in science activities, they will feel better about themselves as science students; they will perform better and will be more highly motivated to explore and experiment.

Your young scientists, too, should be responsible for cleaning up a center so that it is ready to be used by the next scientists. Ask your students or your child for suggestions in compiling a checklist that can be placed in a prominent place within the center.

SCIENCE TOOLS

A science tool can be defined as any object, utensil, item, or instrument that serves as an implement to aid in exploring, experimenting, examining, testing, or measuring.

The tools in this book fall into three categories.

1. *Scientific tools:* commercial microscopes, binoculars, and stethoscopes
2. *Everyday tools:* thermometers, measuring devices, and batteries
3. *Tools that children can construct:* cardboard-tube telescope, binoculars, and rain gauges

Many science tools have interesting origins. Assigning students research projects on the history and origin of tools, as well as on the tools' important contributions to society, will extend the horizons of young scientists.

Don't overlook the inventors themselves. Understanding how and why a tool was invented and the background of the inventor may encourage and motivate future inventors. Studying and comparing the early and present-day tools will help children understand the progress that has been made.

The following is a list of tools that you may want to collect for use with the activities in this book. Most of the tools can be collected from other parents, borrowed, or purchased at garage sales and discount stores. Microscopes, hand lenses, and chemicals are available at science supply stores.

magnets (assorted types and sizes)
materials that magnets can attract and repel
thermometers (assorted types and sizes)

stethoscopes
binoculars and opera glasses
telescopes
kaleidoscopes
mirrors
flashlights
prisms
scales
measuring sticks, cups, and spoons
rain gauge
battery (12-V)
lenses
cameras
magnifying glasses
clear plastic drinking glasses
eyedroppers

THROWAWAYS BECOME SCIENCE AIDS

Many items that are considered "trash" can be recycled into both tools and aids for science projects. Children become more aware of the importance of conservation and recycling by finding several uses for empty containers, cardboard tubes, and other throwaways. Attach a recyclable item to a piece of posterboard and start a running list of uses for that item. Allow children to add their own ideas. Have them "write" rebus notes to take home, write recycling poems, or write and perform skits and puppet plays.

Establish recycling habits within the classroom or your own home. Provide receptacles for scrap paper, broken crayons, "good junk" collage materials, and special items that you want to collect.

Make recycling a part of math by creating bar or picture graphs to represent recyclable items found by the children. Perhaps a contest between student groups or within your family will stimulate a desire to save reusable items. With a little planning, you and your young scientists could involve an entire school—and other parents, too!

The following recyclable items can be collected to use for the activities and experiments in this book:

- cardboard tubes (assorted sizes)
- margarine containers (large and small)
- quart jars
- olive bottles with straight sides
- coffee cans
- milk cartons and plastic jugs
- fast-food take-home buckets
- shoe boxes
- soft-drink bottles
- tennis ball cans or similarly shaped potato chip cans
- paper and cardboard scraps

STORAGE

Store recyclable items near the science centers where they will be readily available. The greater the variety in materials, the better. Items should be clean, sorted, and stored in appropriate containers.

Stress the importance of returning materials and equipment to their proper place.

Apple and orange boxes are invaluable, especially those with hand slots. They can be easily slid in and out from under a work table.

Fast-food take-home buckets and one-gallon plastic milk jugs (with the tops cut off) make good storage containers for small items.

Plastic or cardboard shoe boxes, department-store boxes in a variety of sizes, and egg cartons are also good for storage. These and other flat containers can be stacked to save space.

Be sure to label *all* storage containers.

CHAPTER 2

A TIME FOR LEARNING ABOUT SCIENCE

TIME TO OBSERVE, EXPERIMENT, AND DISCOVER

Any time can be science time. Today's children are growing up in a science-dominated world, and we, as teachers and parents, need to take advantage of every opportunity to introduce and reinforce useful science concepts. Science is all around us, in everything we do; therefore, it is easily integrated into every area of the school curriculum as well as into fun time at home. Spontaneous science will often prove to be the best science of the day. However, one needs to plan, prepare, and provide appropriate materials for the children to work with, at home as well as at school.

Don't rush young scientists. In order to make discoveries, all scientists need time, patience, and opportunities to experiment. Make sure that you allow sufficient time to complete activities, and provide those desiring it an opportunity to repeat their experiments. Some children may wish to continue an activity the next day or for several days. Giving them ample time to explore, investigate, and experiment will aid in developing patience.

Try to provide additional time during the day when the youngsters can work at leisure on a project. Never throw out a child's experiment without permission. Their experiments, no matter how trivial such projects may appear to you, may be important learning experiences for your young scientists. Being too restrictive, having too many rules, and having a program that is overly structured are sure ways of turning kids off.

Allow kids to make and to discover their own mistakes. Failure can often be the motivating factor for determination.

DAILY LIVING DISCOVERIES

In our busy daily lives, we seldom take time to look at the clouds, at drops of water in a sprinkler, or at an insect on a plant. We often don't look at these or other features of our environment because there seems no particular reason to do so. As parents and teachers, however, we should take advantage of every opportunity and encourage children to take time to look and to ask questions. Train yourself to be a good observer and to always show a sincere interest in the observations and discoveries of others. Pausing to wonder and reflect; noticing the beauty of shape, color, pattern, and texture; and sharing ideas and observations with children will help them to develop desirable attitudes and behaviors.

Provide activities that will help children become more aware of their environment. Make a list of questions related to things that your children see every day. Have them write down their answers. If they don't know the answers, give them time to find out for themselves. Here are some suggestions for questions:

1. Is a police badge worn on the left or the right side of the uniform?
2. On which side of a snail's shell do the spirals appear?
3. Is the "off" position on the standard electric wall switch at the top or bottom?
4. How many tines are there on a dinner fork?
5. Where is the closest fire hydrant to the school?

According to W. Ralph Ward, "Life is not so much a matter of discovering something new as it is a matter of rediscovering what has always been present."

TAKE A GOOD LOOK AROUND

Try having small groups explore different sections of the classroom. (Or try this on a rainy day at home.) After a few minutes, have the children write down their individual observations and then compare them with those of the other members of the group. All will be surprised at the things they overlooked.

How accurate a picture can your children draw of the playground?

To test both observation and memory skills, send the

children outside with paper and pencil. Have them make a list of everything that's on the front wall of the classroom.

For an interesting outside-discovery project, children will need to work in small groups. Give each group a piece of heavy string twelve feet long. Have them use it to stake off a section of ground or an area of grass. See what discoveries they can make in this small section of the environment. Magnifying glasses and a few old spoons for digging will stimulate additional interest.

A picture graph could be made, after digging a hole twelve inches square and six inches deep, showing how many rocks, insects, and other items were found.

SENSORY WALKS

Fill your senses; light up your life! A walk around the schoolyard or the neighborhood park can reveal some surprising information as well as many often-overlooked details. A five-minute silent walk will make kids more aware of their individual senses. Set the boundaries for the walk and forbid talking once the children walk out the door. Follow up with an oral discussion. Children will learn to enjoy the world around them by just "looking."

Walks that cause the observer to concentrate on individual senses—a seeing walk, perhaps, or a smelling walk, a feeling walk, a listening walk—will help train children to tune into their surroundings. In addition to identifying what they see, smell, and hear, have them describe how it looked, smelled, and sounded.

For an experience in listening, have children sit outside and keep their eyes (and mouths) closed for three minutes.

Then ask them to write down all the sounds of which they were aware. Practicing the familiar adage "Stop, look, and listen" is a good habit to instill in all beginning scientists.

As a little reminder, can you:

See—not just look?
Listen—not just hear?
Feel—not just touch?
Smell—not just sniff?

PART TWO

SOME BASIC TOOLS OF SCIENCE FOR TEACHING AND LEARNING

CHAPTER 3

THE MAGNIFYING GLASS: A BASIC LEARNING TOOL

Every young scientist needs a good magnifying glass (hand lens). Choose one with a sturdy frame and an easy-to-hold handle. A three-inch shatterproof acrylic lens, 3× to 5× power (magnifies three times or five times), is ideal for children's use.

What is a magnifying glass? It is a lens that makes close objects appear larger. Both sides of the lens curve outward to form a double convex lens. It can give two kinds of images. A glass held close to the page forms a "virtual image." The light rays do not pass through the lens, and the image appears upright and larger than the object. A "real image" is formed when light rays from an object pass through the lens and are focused on the other side; it appears upside down and reversed. Its size depends on the distance of the object from the lens. The following experiences will help acquaint children with the wonders of the magnifying glass.

Materials: For each child, a clear plastic drinking glass, six-inch square of wax paper, eyedropper, pencil, hand lens, lentil seeds, piece of black paper, four-inch square of cardboard, two-inch square of clear plastic wrap, sheet of newspaper, tape.

Experiment #1

1. Stand the pencil up behind the glass. Look through the glass at eye level.
2. Place the pencil in the glass and hold it upright. Look through the glass at the pencil.
3. Fill the glass with water and repeat steps 1 and 2. Observe the changes.
4. Drop the pencil into the glass of water. Look through the glass at the pencil. How does the pencil appear?
5. Have your friends take turns placing their faces near the glass as you look at them through the water.

Thin glass does not change the way objects look when viewed through it. When you look through something thick and curved (such as a glass of water), objects look different. What you see is an illusion caused by the bending of the rays or waves of light (refraction).

Experiment #2

Place the wax paper on top of the newspaper. Using an eyedropper, make a drop of water, about the size of a dime, in the center of the wax paper. Observe at eye level. What shape is the drop? Now look down through the drop at the newsprint. What do you see? Carefully pick up the wax paper. Tilt one end, but don't let the drop roll off. What happens to the drop? Again look through the water at the newsprint. Make another drop directly on the newsprint. What happens? Why?

Experiment #3

Cut a hole the size of a quarter in the center of the cardboard. Tape the plastic wrap over the hole. Lay it on the newsprint. Use the eyedropper to make a large drop of water on the plastic wrap and then look down into it.

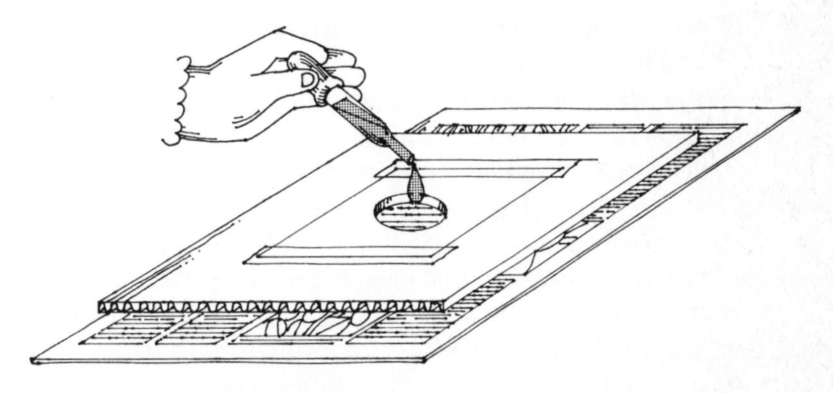

Raise the cardboard carefully and note the changes in the print. You have made a magnifying glass! A curved shape that you can see through is called a *lens.*

Experiment #4

Examine a magnifying glass. How do you think it is made? Lay it on the newspaper. Raise and lower the lens until you find your clearest viewing point. This is called the *focal point,* and it will be different for each person. What happens when you continue to raise the magnifying glass?

A magnifying glass has a curved shape that is thickest in the middle.

Experiment #5

1. Place another magnifying glass on top of yours. Look at the newsprint. Look at your friend.
2. Hold the lens directly in front of your face. Look at an object across the room. Slowly move the lens out to arm's length. Note the changes in the object. At arm's length, it will be reversed and upside down.
3. Repeat step 2, using two magnifying glasses held together.

Additional Experiments

1. Place a drop of water on your magnifying glass and look through it.
2. Stand with your back to a window and hold the magnifying glass in front of your face. Move it about until you see your reflection. Can you see anything behind you?
3. Use your magnifying glass to examine a lentil seed. Separate the two halves of the seed with your fingernail. Compare the shape of the seed with that of your hand lens. The word *lens* comes from the Latin *lentil* because its shape resembles that of a lentil bean. A lentil also has two curved surfaces.
4. Look at a piece of string. What do you see?
5. Shake a few grains of salt onto one side of a piece of black paper. Place a few grains of sugar on the other half. Compare the two. Both are crystals. What shape are the crystals?
6. Look closely at a photograph in your newspaper. The picture is made up of tiny black and white dots. Can you see them?
7. Some other interesting items to view with a hand lens are:
   a. the inside of a mushroom
   b. the cut end of a piece of celery
   c. spores on the back side of a fern
   d. grains in a piece of wood
   e. threads in a piece of nylon hose
   f. your fingertips

g. grains of sand
h. sponges (both ocean and synthetic)
i. the torn edge of a paper towel
j. the roots of a blade of grass
k. veins in a leaf.

A WATER SCOPE

Materials: cardboard paint bucket or a fast-food take-home bucket, clear plastic wrap, a large rubber band, water, some small objects (paper clips, pencils, toys).

Activity:

1. Cut a hole in the side of the bucket.
2. Stretch the plastic over the top of the bucket and secure it with the rubber band.
3. Slowly pour one-fourth cup (or less) of water onto the plastic wrap.
4. Place objects in the bucket through the side hole.
5. Look through the top of the water scope.

CHAPTER 4

RELATIVES OF THE MAGNIFYING GLASS

A collection of devices that contain lenses may contain more objects than you think! What might your collection encompass? Suggestions: eyeglasses (reading, bifocal, trifocal, contact lenses), cameras (instant-print, fixed focus, single-lens reflex, twin-lens reflex, motion picture, telephoto and wide-angle lenses), binoculars and opera glasses, telescopes, and microscopes.

Discuss with your young scientist the various tools and discover the purpose of each type of lens. Compare the two basic types of simple lenses: convex (those that curve outward) and concave (those that are hollow and curve inward like a cave). All lenses have at least one curved side that bends the light as it passes through. Reexamine the magnifying glass.

Look into the eye of a friend. Light enters the eye through the pupil (the iris actually lets the light in). Have your friend close his or her eyes and then ask if he or she can see you. Take an instant-print photo. Place your finger over the lens and take a second picture. What happened? Compare the camera's lens with the eye.

Both the eye lens and the camera lens let the light in. The human eye sees still and moving pictures—with or without color. It automatically adjusts to speed, distance, and brightness of light. Can a camera do this?

TELESCOPES

If you want to magnify an object that is far away, use a telescope. A spy glass is a small telescope. A simple telescope can be made by following the directions below.

Materials: one thick and one thin convex lens, masking tape, two cardboard tubes (one about fourteen inches long and one about eight inches long that will fit snugly into the longer one).

Activity: Slide the small tube halfway into the large tube. Tape the thick lens to the open end of the small tube and the thin lens to the open end of the large tube. Slide the small tube in and out as you focus on an object.

Look out the window or down the street. Without moving your head, name the objects you see. Now do the same with your telescope. How is your view different? (Objects will appear upside down. In order to make the image appear right side up, a double convex lens must be placed between the two existing lenses. Some telescopes also contain mirrors.)

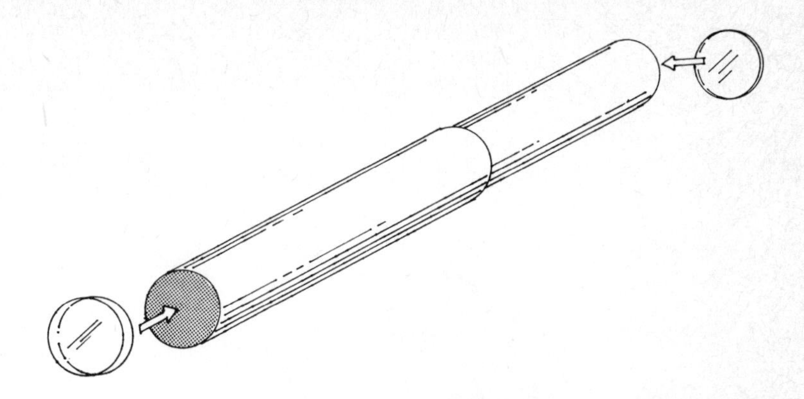

The telescope should cause you to "zero in" on an object or a small area while cutting off your peripheral (side) vision.

The periscope works on the same basic principle as a telescope. However, it also uses reflecting mirrors. See Chapter 7.

BINOCULARS

Binoculars, which make distant objects look larger, allow you to use both eyes for viewing. By making the binoculars described here, you will better understand how they work.

Materials: four cardboard tubes (about five inches long), four paper clips, masking tape, two thick and two thin convex lenses.

Activity: Place two tubes side by side. Clip the tubes together. Then wrap tape around the two tubes to hold them securely together. Do the same with the two other tubes. Tape the lenses in the ends of one set of tubes. The thick lens should be at the viewing end (the end next to the eyes).

Use the tubes without lenses to look at a distant object. Now look at the same object with your binoculars. What dif-

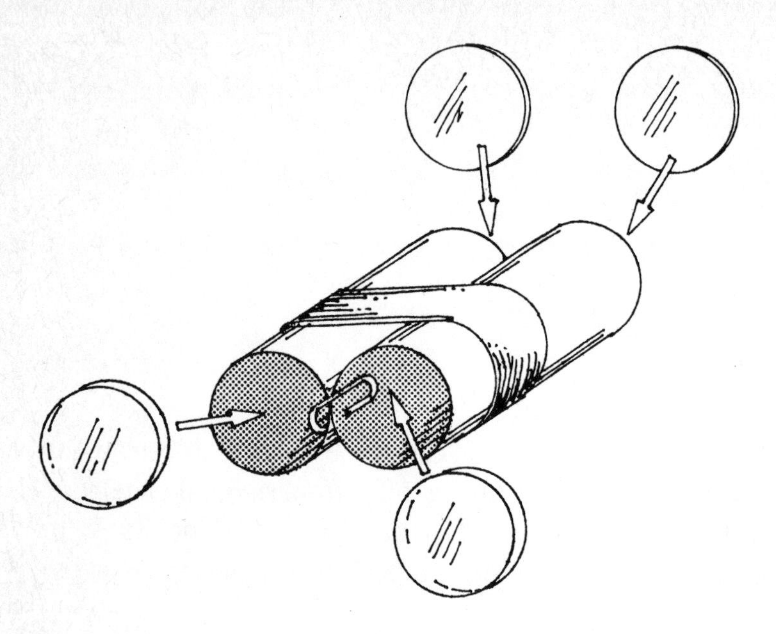

ference do you observe? (The image should appear upside down and the right and left sides reversed. Most binoculars contain a set of prisms in each tube that inverts and reverses the image.)

Close one eye and repeat the experiment.

A VIEW THROUGH THE MICROSCOPE

A microscope magnifies tiny objects that the eye cannot see. It is much stronger than a magnifying glass. A 10× lens will magnify an object (specimen) ten times its normal size.

Materials: microscope, commercial set of prepared slides, several empty slides, tweezers, eyedropper. Spores from the back of a fern plant, stagnant rain or pond water, aphids from a flower bud or a leaf, pollen, a leg or eye from a dead insect, grains of salt, a fingernail clipping, and a strand of hair are all interesting specimens for observation.

Activities: Use the tweezers or eyedropper to place a small specimen on a clean slide. Set a second clean slide on top. Place the slide in position and adjust the microscope. Refer to your microscope manual for proper use and adjustments.

Look closely at a small specimen. Observe the same specimen with a magnifying glass. Now place it under a microscope. Compare the three views to see what details you missed in your first two observations.

Looking at specimens through a microscope will help you understand its power and how the lenses work. It will also enable you to see objects that you cannot see with your eyes alone.

AND WHAT ABOUT STETHOSCOPES?

The tools described have shown us how lenses can focus light waves. In a similar way, stethoscopes can be used to focus and magnify sounds made by the heart, lungs, and other organs.

Materials: stethoscope and a timepiece with a second hand. See "Think and Do" #1.

Activities: Place your ear against a friend's chest and listen to the heartbeat. Use the stethoscope and listen again. Did the stethoscope "magnify" the sounds made by the heart?

Use the second hand on the timepiece to time the heartbeats of a friend while at rest. While at rest, the average adult heart beats about seventy-two times a minute. A child's heart beats slightly faster. Have your friend jump up and down ten times. Now time the heartbeat again. What did you discover?

Listen to the heart and lungs while your friend breathes normally. Then ask your friend to breathe deeply, inhaling through the nose and exhaling through the mouth while you listen. What comparisons can you make?

Time your friend's normal breathing while at rest. Then have your friend jump up and down ten times and time the breathing again. What does this comparison tell you? The average adult breathes about fourteen times a minute. Your friend's breathing rate will be more rapid as the breathing rate decreases with age. For example, during the first year of life the breathing rate is about forty-four times a minute, and at age five years, twenty-six per minute. The rate continues to reduce and reaches the adult average between the ages of twenty and twenty-five.

OTHER RELATIVES OF THE SCOPE FAMILY

A stereoscope is another interesting device containing lenses. What do you think it is? Check the dictionary or encyclopedia to see if you are right. (You may find one in a library or an antique store.)

A kaleidoscope, another object for viewing, is presented in Chapter 7.

CHAPTER 5

MATH AS A TOOL

For an innovative approach to predicting, estimating, measuring and graphing, try these "math ideas kids eat up." "Yucky arithmetic" becomes more palatable and lots more fun in the kitchen.

POPCORN MATH

Materials: electric popcorn popper, popcorn, ten-foot-square plastic sheet (painter's dropcloth), yard/meter stick or measuring tape, eight three-inch by one-inch strips of card-board, hole puncher, thumbtacks, felt-tip pens. See "Think and Do" #2.

Activities: Use the felt-tip pen and measuring stick to divide the plastic sheet into eight equal triangle-shaped spaces. Set the popcorn popper on a tray and place it in the center of the sheet. Assign each person a section, reserving one for the supervising adult and the electrical cord.

Choose three spots, within your space, where you predict the greatest number of kernels will land when the corn is popping and the lid is removed. Using a felt-tip pen, make dots to represent the places.

Make a drawing compass by punching a hole in one end of the cardboard strip. The hole should be centered on the width of the strip and ½-inch from the end of the cardboard strip. Measure ½-inch from the other end of the strip and press a thumbtack through the cardboard. Place the thumbtack on one of the dots. Hold it firmly. Place a felt-tip pen in the punched hole. Use the pen to move the cardboard strip around until you have drawn a circle. Do the same with the other two dots. (This may require some practice beforehand.) Write "1," "2," and "5" in the circles to represent point value.

Estimate (and record) the total number of kernels that will land in your space. Estimate and record how many will land in each circle.

Preheat the popper, add the popcorn, and cover the popper. Give instructions to stay away from the plastic sheet and not to touch the popped corn. When the corn starts to pop, remove the lid and watch the fun.

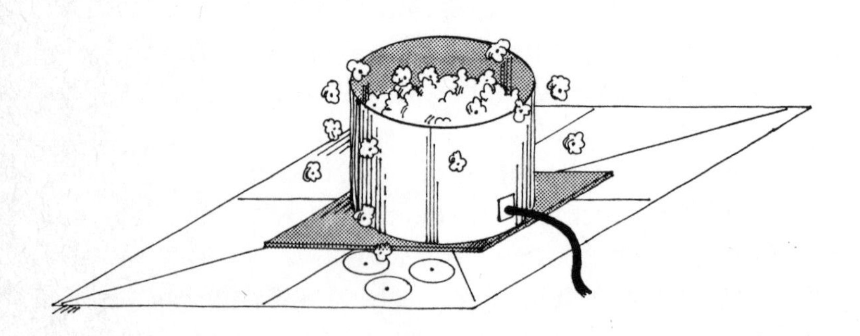

After the last kernel has popped, remove the popper. Count the kernels in your space. Count the number of kernels in each circle and multiply by the appropriate point value ("1," "3," and "5"). Figure your total point value. Record the information and compare it with your original estimates.

Which kernel popped the farthest? Measure and record the distance.

Eat and enjoy the popcorn that landed in your space.

Make a group graph showing the number of kernels that landed in each section.

Make a group "distance" graph.

PEANUT MATH

Materials (for each group of four): a pint jar of peanuts (in the shell), four paper towels, scales. See "Think and Do" #3.

Activities:

1. Each person should estimate and record the number of peanuts in the jar.
2. Pour the peanuts onto the four paper towels and have each person count his or her pile.
3. Total the four piles. How close was your estimate?
4. Put all the peanuts in one pile. Divide the total amount of peanuts by four and have each person count out his or her equal portion. Place your peanuts on your paper towel. Record the number you have.
5. Estimate and record the number of whole peanuts inside the shells.
6. Shell, count, and record. Compare with your estimate.

7. Estimate the weight of the shells and the weight of the peanuts. Record.
8. Weigh the shells and peanuts separately. Record.
9. Set up a math example to show the comparison of shells to peanuts (e.g., addition, subtraction, fractions, or percentage).
10. Eat and enjoy the peanuts.
11. Save the shells. Ask the children for suggestions on how the shells could be recycled (collages, finger-cap puppets, boats, other art ideas).

POTATO MATH

Materials (for groups of eight): one small potato cooked with the skin on, three pounds of powdered sugar, eight forks, eight measuring tablespoons, eight extra spoons for stirring, eight margarine tubs. (Optional: corn starch, cooked sweet potato, and brown and granulated sugar.) See "Think and Do" #4.

Activities: Peel and divide the potato into eight portions.

1. Place a potato piece in a tub and mash until smooth.
2. Predict what will happen if you add powdered sugar to the mashed potato.
3. Add one level tablespoon of powdered sugar and mix well. What changes do you see taking place? (The mixture should be liquefying. If it isn't, add more sugar.)
4. Estimate and record the number of tablespoons of powdered sugar it will take to form the mixture into a firm ball.
5. Add another tablespoon of powdered sugar and mix well. Continue this process, recording the number added, until you have a firm, nonsticky ball.

6. How many tablespoons of sugar did it take? Compare with your estimate.
7. Eat your potato candy. Can you taste the potato? Describe how the candy tastes.
8. Predict what would happen if you added each of the following to a cooked, mashed potato cube.
   a. granulated sugar
   b. brown sugar
   c. cornstarch
9. Predict what would happen if you added powdered sugar to a piece of cooked, mashed sweet potato. If you added:
   a. granulated sugar
   b. brown sugar
   c. cornstarch

If possible, do the experiments in #8 and #9 and find out if your predictions are accurate.

CHAPTER 6

WEIGHT AND MEASURE TOOLS

Spring scales (bathroom, baby, postage, and diet scales) will provide extended opportunities for making discoveries. Collect as many different kinds of weighing scales as possible. The oldest type is the balance scale. The equal-arm balance scale consists of a basket tied to each end of a horizontal beam. A steelyard balance scale is the type used in doctor's offices; a weight is moved along the beam to balance the person's weight.

Provide a variety of material and objects for weighing and allow ample time for experimentation.

HOW TO MAKE A BALANCE SCALE

Materials: a 12″ × 12″ × 1″ board and a 2″ × 2″ x 12″ board, one 12-inch-long lath (¼″ × 1½″) nails, string, hand wood drill, two large margarine containers, two plastic zip-top bags, two wire hooks, measuring cups and spoons.

Activities: Punch or drill a small hole one inch from each end of the lath and insert the wire hooks. Drill another hole in the center of the lath. Place a nail, smaller than the hole, through the hole and nail it loosely to one end of the 2″ × 2″ board. Be sure to center it. The arm (lath) should move easily. Nail the other end of the two-inch board vertically to the center of the 12″ × 12″ board.

Punch four holes, at equal distances, around the top of each margarine container. Cut eight six-inch lengths of string. Tie one through each hole. Tie the four loose ends together and hang one tub on each hook. If you have measured correctly, your scale should balance. When balanced, mark an arrow on the top center of the vertical board and a matching dot on the horizontal beam.

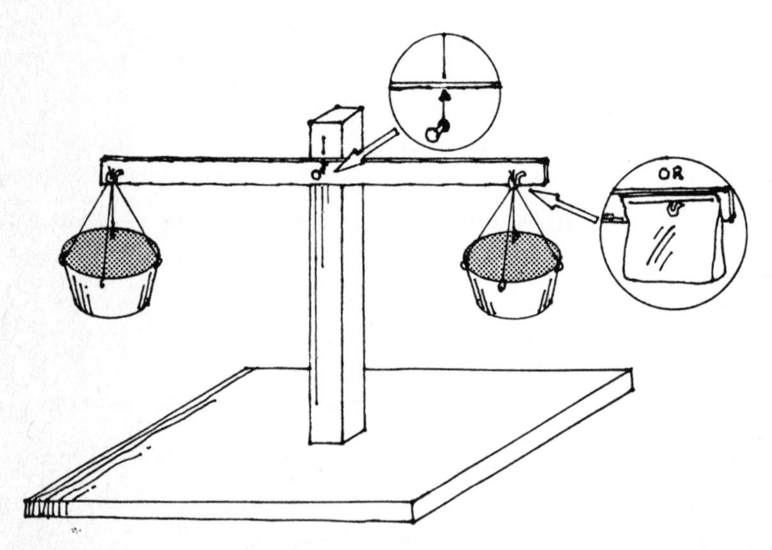

You can use the plastic bags in place of the margarine tubs by punching a hole in the top center of each bag.

Use your balance scale to determine which is heavier, equal amounts of:

1. salt or dry cereal
2. rice or flour
3. water or flour
4. "plain" (liquid) water or frozen water
5. a lemon or a rock of equal size.

Substitute similar items or experiment with any available material.

Can you weigh air? Attach an open plastic bag to one hook. Blow up a second bag, seal it, and hang it on the other hook. Does air have weight?

WEIGHTS

Materials: sand, salt or dried beans, small heavy plastic bags with ties, felt-tip markers, diet or postage scale.

Activities: Place varying amounts of material into the plastic bags so that each bag will equal a different number of ounces. Tie at the top and mark the weight on each.

Place a bag in one basket. Add material to the other basket until the scale balances.

MEASURING TIME

Assemble as many objects for measuring time as possible—spring-driven, weight-driven, electric and digital clocks; one-

and three-minute sand timers; a sixty-minute spring kitchen timer; and pocket, wrist, and stop watches. Experiment to find how they are similar and how they are different.

Activities: Synchronize all the clocks and watches. Check at the end of the day to see if any of the clocks have gained or lost time. Which ones are the most accurate?

How long is a minute? Have everyone close their eyes. Turn a minute timer upside down. Have the children raise their hand when they think a minute is up.

Ask your kids to raise their hand if they think they can run in place for one minute. Time their running. Although most thought they could, studies have shown that 75 percent of elementary school-age children cannot run in place for a full minute.

SHADOW CLOCK

Materials: pencil and a paper plate.

Activities: The first method devised to tell time was to observe and mark the movement of one's own shadow or the shadow of a permanent vertical object.

On a sunny day, you can become your own shadow clock. Check the clock for the exact hour and stand with your back to the sun. Mark your spot by placing a rock at your feet. Have someone make a line to mark the top of your head's shadow. Record the hour next to the line. Repeat every hour on the hour. What happens to your shadow?

Make a simple shadow clock by pushing a pencil halfway through the center of a paper plate. Poke the pencil into the ground and mark the position of the shadow. Repeat every hour on the hour.

WATER CLOCKS

A water clock (clepsydra) was another early method of keeping track of time. A small trickle of water was used to either empty or fill a container. They were used by both the Chinese and Greeks. Experiment with water clocks by making the two types described here.

Materials: for (1), a large and a small margarine container. For (2), a large margarine container and a large jar with straight sides (you should be able to set the bottom of the margarine tub just inside the mouth of the jar). And for both, a very small nail, a felt-tip pen, water, and a clock with a second hand.

Activities:

1. Use the nail to punch a small hole in the bottom of the small container. Fill the larger container with water and set the smaller one on top of the water.

   When the small tub fills with water and sinks, this will equal one interval of time. Time this process to determine the duration of the interval. If the tub sinks in less than a minute, make a smaller hole in a new tub and repeat the process.
2. Punch a tiny hole in the bottom center of the margarine container. Make a mark around the inside near the top. Set the container on top of the jar. Fill the margarine container to the mark with water.

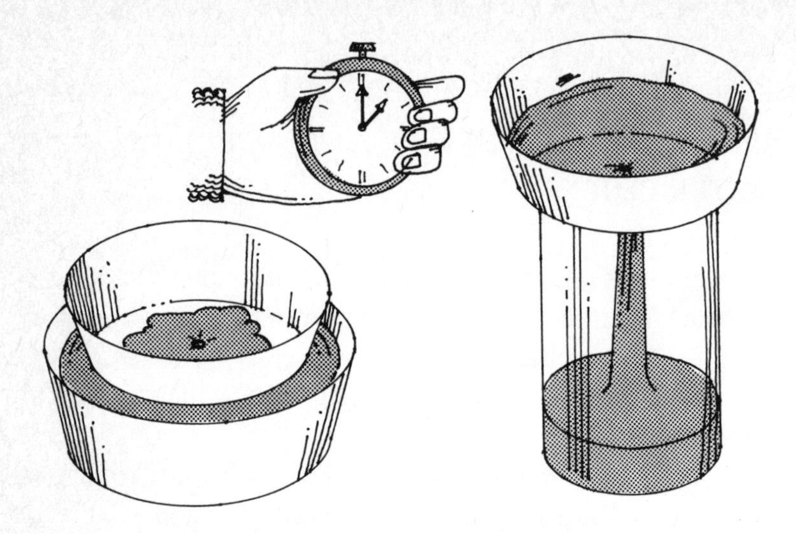

When the top container is empty, one interval of time has passed. Start timing as the water starts to drip. At the end of each minute, mark the level of water on the outside of the jar.

SUNDIAL

You can use a sundial to tell the time by measuring the angle of a shadow cast by the sun.

Materials: one 8″ × 8″ × 1″ board, a pencil-size wooden dowel, glue, a magnetic compass, a felt-tip pen.

Activity: Draw a line down the center of the board. Glue the dowel upright in the center of the board. Set the board in direct sunlight away from trees and buildings. Using a magnetic compass, adjust the board so the line faces north and south. Mark the location of the shadow cast by the dowel (when used for this purpose, it's called a gnomon) on the board. Start your marking on the hour and continue making a mark for each hour.

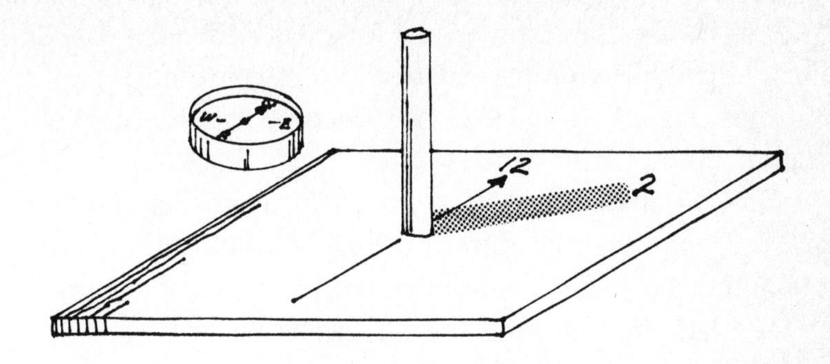

Set an hour timer to remind you to check the sundial. Write the hours on the board of the sundial.

During the next few days, your sundial will remain fairly accurate, and you will be able to calculate the time within ten to fifteen minutes. After about a week, you will need to use your compass to change the position of the board. Why is this necessary? Why do you think a sundial would not be a practical or accurate way of telling time?

A more accurate sundial is one made with a slanted gnomon of metal or wood. The upper edge slants upward at an angle equal to the latitude of the geographical location of your sundial (the angular distance from your location to the equator). For this reason, and since your magnetic compass does not point true north, several adjustments would need to be made. You may want to research this type of sundial. However, the sundial you have made will help you to understand in general how a sundial works.

SAND TIMER

An hourglass, another device that measures time, consists of two containers with tiny openings. One is filled with fine, dry sand and then they are joined together at the openings. When all the sand drains out into the other container, one hour has elapsed. The hourglass is then turned over to begin another hour.

Similar instruments are made on a smaller scale to measure one or three minutes. Egg and game timers are usually made of plastic and filled with salt. A minute timer can be made by following the directions given here.

Materials: a cone-shaped paper cup, straight pin or pointed scissors, fine sand or salt, a pint jar, a timepiece with a second hand.

Activity: Use the straight pin to punch a tiny hole in the bottom of the paper cup. Place the cup on top of the jar so that it extends above the top. Fill the cup with sand. To time the sand, mark the outside of the jar when one minute has elapsed. Remove the remaining sand from the cup. Pour the sand from the jar back into the paper cup. You now have a one-minute timer. Salt may require a smaller hole as it may flow too quickly. By using bird gravel, you can cut a larger hole in the cup. Experiment to see if you can produce a two- and a three-minute timer.

RAIN GAUGE

A rain gauge is an instrument used to measure the amount of rain that falls in one place during a given period of time. Rainfall is measured in inches, or in millimeters when a metric gauge is used.

Materials: a straight-sided can (a one-pound coffee can

will do), a tall, straight-sided glass jar (olive jar), measuring stick, felt-tip pen, water.

Activity: Pour water into the can to a depth of two inches. Then pour the water into the bottle. Mark the depth of water on the outside of the bottle. Pour out the water. Use your measuring stick to divide the space below the mark into twenty equal spaces. Each mark will represent one-tenth of an inch.

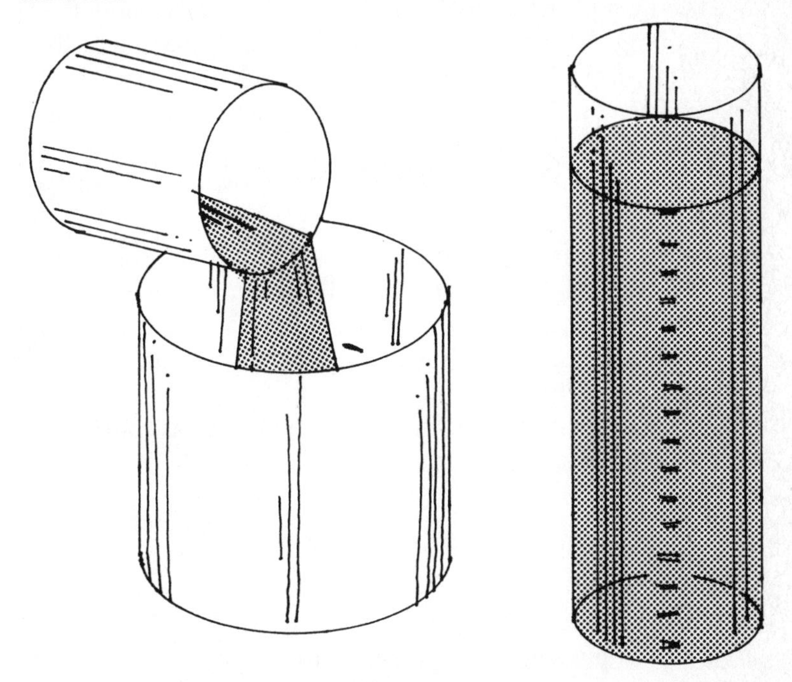

Find a safe place, away from trees and buildings, to set your can. Record the time it starts and stops raining. After the rain, pour the water from the can into the jar. By reading your gauge, you can determine the amount of rain that fell within that period of time.

MEASURING SNOW AND ICICLES

Materials: a straight-sided can, ten to twelve inches tall (a two-pound coffee can), measuring stick, liquid measuring cup, snow and icicles.

Activities:

1. Is an inch of snow equal to an inch of rain? You can find out by filling the can with snow (don't pack it down) and then setting it in a warm place until it melts. Now measure to see how many inches of water are in the can. What did you discover?

   It takes about *ten to twelve inches* of snow to equal an inch of rain.
2. Break off an icicle and measure its length. Place it in a container and let it melt. Now pour the water into a measuring cup. What does this experiment tell you?

   Your ______-inch long icicle equals ______ cup(s) of water.

   Will every melted icicle the same length as yours equal the same amount of water? Why?

MEASURING UP

Materials: measuring sticks, tape measure, catalogue or newspaper advertisement for carpeting.

Activities:

1. Using the customary system, measure your room for a new carpet. Find the square feet by multiplying the length times the width. Select a carpet from the catalogue and figure the cost of carpet for your room. (Carpet may be sold by the

square yard; if so, divide the area by nine to find the number of square yards needed.)

2. Use a tape measure to measure the circumference of a pole, barrel, or large ball.
3. Using the metric system, measure the length of a building on the outside or a basketball court.

If possible, provide the kids with a tire gauge, a tire tube and a hand pump, a speedometer, and an odometer for experimentation.

Examine the gauges on utility (water, gas, and electric) meters.

CHAPTER 7

LIGHT AS A TOOL

If you set up a visual center, children will enjoy this special opportunity to explore and discover.

Materials: prisms, assorted mirrors, magnifying mirror, flashlights, colored cellophane, colored glass bottles, different lengths and sizes of cardboard tubes, eyeglasses (plain and tinted), diving mask, goggles, View Master, translucent and clear plastic lids, opaque objects.

Discuss and compare items that are:

Transparent: allows all the light rays to pass through. Objects can be seen clearly through a transparent surface.

Translucent: allows some light rays to pass through. Objects can be seen, but not clearly, through a translucent surface.

Opaque: allows none of the light rays to pass through. Objects cannot be seen at all through an opaque surface.

SHADES OF SCIENCE

Materials: red and white liquid tempera paint, eyedroppers, small paper plates, plastic spoons, margarine tubs, small paint brushes or cotton swabs.

Activities: A pair of normal human eyes can, under the best possible viewing conditions, distinguish from 8 to 10 million differences and gradations in color.

How many shades of red do you think there are?

Pour a small amount of red paint into a margarine tub. On the paper plate, paint a narrow stripe from the rim to the center of the plate. Add *one* drop of white paint to the tub, mix well using a spoon, and then make another stripe bordering the first one. (Be sure to rinse the brush thoroughly after painting each stripe.) Repeat, adding *one* drop of white paint at a time to the existing red paint, mixing, and then making another stripe. Continue around the entire rim of the plate. How many different shades did you make? How could you make more shades of red?

COLOR CHANGE

Materials: lids from margarine containers or coffee cans, colored cellophane, tape, scissors, rubber bands, two cardboard tubes. (One tube must be longer and should fit inside the other tube.)

Activities:

1. Cut out the center of the lids, leaving a one-inch-wide rim. Tape a piece of cellophane over the opening. Make several in different colors. Experiment by placing the lids over and un-

der one another to see how many different color variations you can create.

2. Cut out a square of blue cellophane and place it over one end of the larger tube. Secure with a rubber band. Cut out another square of yellow cellophane and cover one end of the smaller tube. Secure with a rubber band. Place the small tube inside the large tube. Push it toward the blue cellophane and watch the colors change. Experiment by replacing the pieces of cellophane with pieces of other colors.

SPECTRUMS, PRISMS, AND RAINBOWS

Prisms are transparent and can be used to reflect light rays, refract (bend) them, or separate their colors. Prisms are used in binoculars, periscopes, and other scientific instruments. Some cameras also contain mirrors and prism lenses. When white light passes through a prism, it is broken up into bands of seven colors called a spectrum.

Activities: Experiment with prisms on a bright, sunny day. The seven colors of the spectrum in their proper order are red, orange, yellow, green, blue, indigo, and violet. Red is the longest wave (bent the least), and violet is the shortest (bent the most).

Clear opals, diamonds, glass beads, prisms from chandeliers, and cut glass objects will also separate light rays into a spectrum.

MAKING RAINBOWS

A rainbow is made when raindrops refract sunlight, breaking the white light into the seven colors of the spectrum and then reflecting it back in the direction from which it came. Although the individual colors are sometimes hard to recognize, all seven colors are present, and they are always in the same order. (You can recall the correct order of the rainbow by associating the names of the color with each letter in the name of a friend named Roy G. Biv.) Red is on the top of the rainbow. If you see a double rainbow in the sky, the colors in the bottom one, a reflection from the top rainbow, will be reversed.

We see the rainbow as an arch, but it is actually a circle that is partly hidden by the horizon. Rainbows can often be seen in the sprays from fountains, waterfalls, and lawn sprinklers.

Here are two ways to make a rainbow when there is no rain.

Materials: a piece of white cardboard, clear plastic glass, small mirror, water.

Activities:

1. Fill the glass halfway with water. Place it in bright sunlight. Place the mirror in the water at an angle so that the sunlight strikes it. Hold the cardboard about two feet from the glass and to one side. Adjust the mirror so the sunlight will reflect from the mirror and form a rainbow on the cardboard.
2. Fill the glass with water. Place the glass on the edge of a table so that sunlight shines through the water. Place the cardboard on the floor. Slowly pull the glass toward the edge of the table while adjusting the cardboard so that the sunlight shines through the water, forming a rainbow on the cardboard.

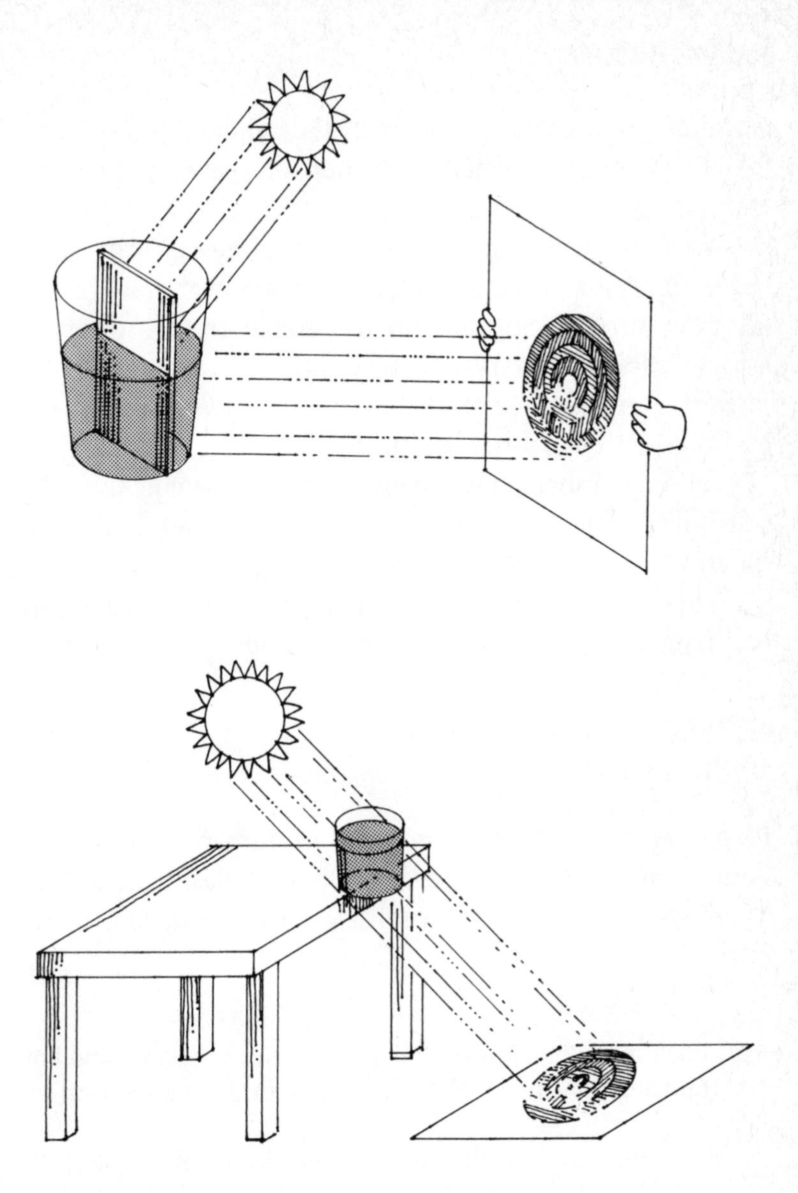

RAINBOW ART

Materials: old, broken crayons in the rainbow colors, hand food grater or hand pencil sharpener, wax paper, yarn, warm iron.

Activity: Grate the crayons onto pieces of wax paper. (Keep the colors separate.) Place another sheet of wax paper on a flat surface. Sprinkle on the grated crayon to form a rainbow. Remember Roy G. Biv. Since the red arch is the longest, remember to start it at the top of your paper. Add each correct color underneath the preceding one. Carefully place a second sheet of wax paper on top. Press quickly with a warm iron. The colors will melt into each other and appear much the same as we see the rainbow. Trim the edges of the wax paper and punch a hole in the top. Attach a piece of yarn and hang your rainbow in a sunny window.

MIRROR AND FLASHLIGHT EXPERIMENTS

Materials: assemble an assortment of mirrors (wall, pocket, hand, and magnifying), several different flashlights, large, shiny spoons, and several front pages from your local newspaper.

Activities:

1. Look into a hand mirror (held with your left hand) and hold up your right hand. Note how a mirror gives a reversed image.

   Now stand with your back to a large mirror (keep the hand mirror in your left hand and your right hand up). Look into the hand mirror. Does your right hand look different? What can you see behind you?

2. Stand in front of a mirror holding up a newspaper. Can you read the headlines? Turn your back to the mirror. Hold another mirror in one hand and the newspaper in the other. Look through the hand mirror and adjust the newspaper so you are reading through the rear mirror. Can you read the headlines?
3. Hold a magnifying mirror in one hand and a regular mirror in the other. Compare your reflections. Slowly push each mirror away from your face. What happens? Do both mirrors have a flat surface? You can find out by placing the mirrors on a level surface, reflecting side up, and then placing a pencil across the center of each mirror.
4. Hold a mirror so that sunlight shines on it at an angle. Reflect the light on the wall or ceiling. Move the mirror about. Try this while holding a mirror in each hand.
5. Can you think of other places where you could see your reflection? (Try metal appliances; chrome bowls, windows, lakes, and puddles.)
6. Examine a large polished spoon. What observations can you make? Look into the bowl of the spoon. How do you look? Hold the spoon sideways and look into it. Turn the spoon over and look at your image. Compare your images in the spoon with the one in a flat mirror.
7. Take apart and examine different kinds of flashlights. Note the positive (+) on the top side of the flashlight battery and the negative (−) on the bottom side.

   Will the bulb light using only one battery?

   Between the battery and the bulb, you will see a funnel-shaped object—a small curved mirror. Experiment and see if you can discover the purpose of this mirror.
8. Place a mirror on the floor. Stand back and direct the

flashlight beam onto the mirror at an angle. The light rays will bounce off the mirror and be reflected onto the wall.

MAKE A CARDBOARD KALEIDOSCOPE

A kaleidoscope can produce a fascinating array of shapes and colors. The basic design consists of a tube containing two mirrors set facing each other at angles of about 45 degrees.

Materials: a heavy cardboard tube about three-and-one-half inches in diameter and nine inches long, two mirrors (three inches by nine inches), scraps of heavyweight construction paper (black and assorted colors), a four-inch-square piece of clear acetate, a small, translucent margarine container lid, 5″ × 12″ piece of heavy cardboard, masking tape, scissors, white liquid glue.

Activity: "Hinge" the two mirrors with a long strip of tape. Slide them into the tube so the untaped edges touch the sides of the tube. Glue in place. Cut a circle of clear acetate to fit one end of the tube. Glue it to the end. Cut a half-inch strip of cardboard long enough to fit around the tube. Wrap it around the same end of the tube to form a rim. About one-quarter inch should extend above the tube. Glue in place.

Cut out two two-inch squares of black paper. Fold the paper in half and cut out a figure such as a geometric shape, person, or tree from each paper. The figures need not be alike. Lay the figures on the plastic circle. Place the lid on top and tape it into place.

Turn the tube upside down. Cut out a circle of cardboard to fit the end of the tube. Now cut a circle about the size of a penny in the center of the cardboard. Glue the cardboard to the end of the tube.

Look through the viewing end. Slowly rotate the tube and watch the numerous combinations of patterns emerge. Viewing can sometimes be enhanced by looking toward a window or other light source.

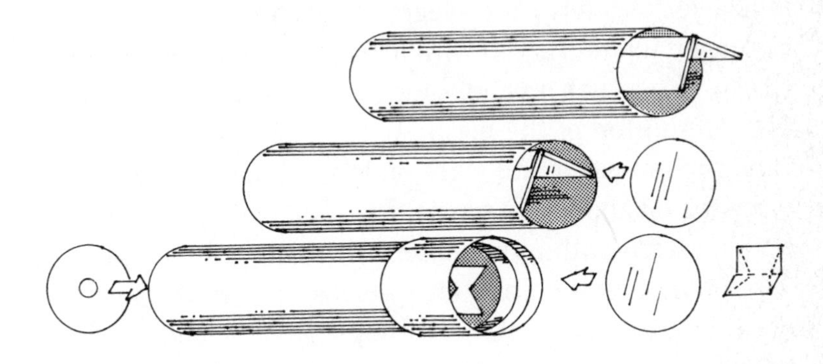

Remove the tape and the lid. Replace the black figures with your own colorful creations.

The tube can be painted or covered with colorful adhesive-backed paper.

When using a tube of a different size, adjust the mirror size accordingly.

TENNIS BALL CAN KALEIDOSCOPE

Materials: a tennis ball or similar type can, 10″ × 5″ piece of heavy-grade silver reflective plastic with a mirror-like finish (known as Mylar) that comes in a roll and is available at hardware and building supply stores, clear plastic wrap, rubber band, twelve-inch piece of cord, scraps of heavyweight construction paper, white liquid glue, scissors, masking tape, hammer and nail. Optional: 4″ × 4″ piece of clear acetate.

Activity: Remove the lid from the can. Trim the piece of Mylar so that it is half an inch shorter than the tall can. Crease the reflective plastic down the middle lengthwise and insert it into the can. It should fit snugly so that the two reflective sides form a 45-degree angle and face each other. Cover the top of the can with a piece of plastic wrap and secure it with a rubber band. Glue the cord along the top rim of the can, on top of the plastic wrap. Cut small figures from construction paper and place on top of the plastic wrap. Replace the lid and tape it to the can. (The figures must be able to move freely within the space.) Use a nail to punch a hole in the opposite end. Point your kaleidoscope toward a window or other light source. Look through the hole and slowly rotate the can as you observe the changing patterns.

If you substitute a circle of clear acetate for the plastic wrap, your kaleidoscope will be more durable. Cut a circle of acetate to fit the open end of the can. Glue it to the top of the can. Glue the cord to the acetate. Complete as above.

VIEWING BOX (PERISCOPE)

Materials: a long, narrow shoe box, two mirrors (approximately 3″ × 6″), masking tape, knife.

Activity: Remove the lid from the shoe box. Set the box on end with one side facing you. Cut out a window about the size of a quarter one inch up from the bottom of the box. Place one mirror in the box, face up. (One three-inch end should rest in the corner of the box on the window side; the other three-inch end should slant at about a 45-degree angle

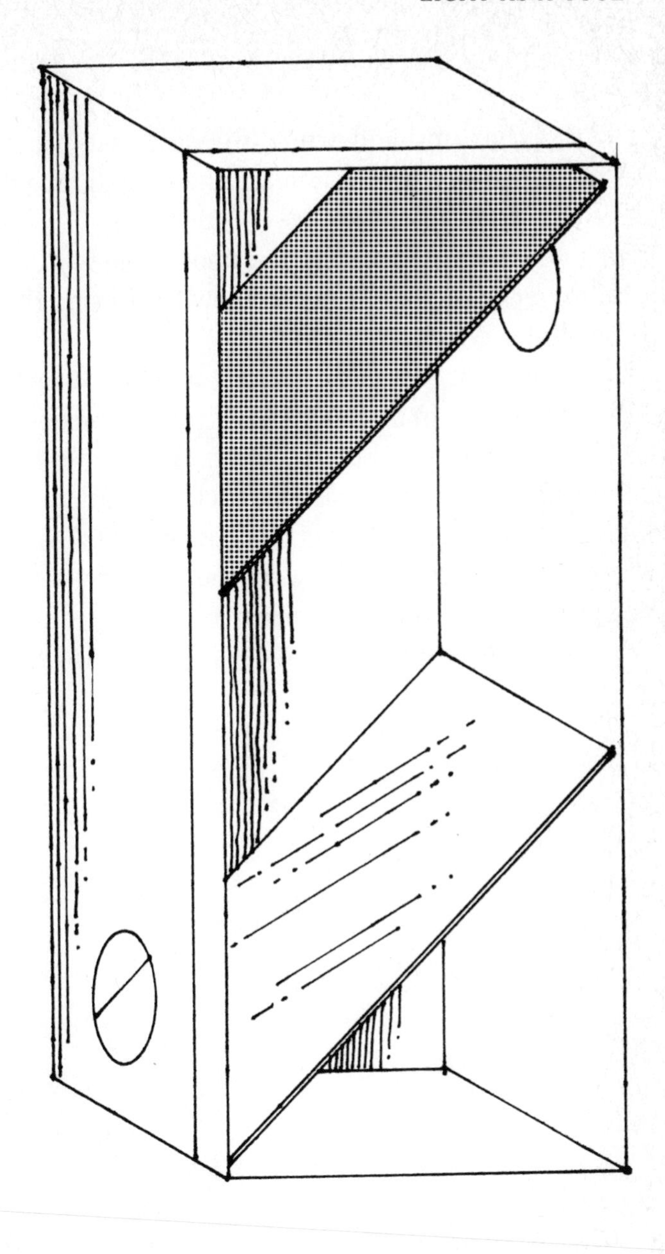

and rest against the back of the box.) Tape the mirror in place.

Turn the box upside down so the other side faces you. Repeat the directions above. Replace the lid and tape it into place.

Place the periscope so that the upper hole is just above the window ledge. Look out the window by peering through the lower hole. What can you see?

Try using your periscope to look around corners. (Although light travels in a straight line, your periscope's mirrors let the light travel around corners.)

Can you read the words on a chalkboard or a sign through your periscope? Why is this possible? (The first mirror reversed the writing, but the second mirror reversed the images produced by the first mirror.)

Periscopes may also contain lenses and prisms.

CHAPTER 8

TEMPERATURE AS A TOOL

Children can learn about the measurement of temperature by experimenting with a variety of different types of thermometers. Collect as many kinds as possible—medical (oral), meat, candy, weather, and the dial types used in ovens and refrigerators. Compare and discuss the similarities and the differences.

You will need several weather thermometers—both Fahrenheit and Celsius—for the following experiments. One type, the double-scale, shows both. Digital thermometers are also available.

The most common liquids found in thermometers are mercury (it looks silver) and alcohol (usually colored red). **Caution:** Mercury is extremely poisonous and can be absorbed into the blood through the skin.

WEATHER THERMOMETERS

Materials: for each group: thermometer, hot water, crushed ice, three clear plastic cups.

Activities:

1. Have children observe the mercury rise and fall in the thermometer by alternately placing and removing their thumb from its bulb.
2. Put the thermometer in a glass of ice water and then in a glass of warm water. Note the changes.
3. Fill one cup with crushed ice, one with tepid water, and one with hot water. Place your finger in each and estimate the temperature. Place one thermometer in each glass and compare the temperature with your estimates.

FAHRENHEIT VERSUS CELSIUS

Materials: Fahrenheit and Celsius thermometers, shallow pan of boiling water, two clear plastic cups, ice, salt.

Activities: On a Fahrenheit scale, 32 degrees is the freezing point of water, and 212 degrees is its boiling point. On a Celsius scale, 0 degrees is the freezing point of water, and 100 degrees is its boiling point.

Children can discover the difference in Fahrenheit (F) and Celsius, or centigrade (C), by performing the following experiments.

Place each type thermometer into a glass of crushed ice for three minutes. Observe. Place both thermometers into a shallow pan of boiling water (make sure that the Fahrenheit thermometer will register 212 degrees) for three minutes.

You may want to do a comparison using a double-scale thermometer.

Do the experiments that will prove the following:

	°C	°F
Freezing point of pure water:	0°	32°
Freezing point of ice, salt, and water:	–18°	0°
Body temperature (place under arm):	37°	98.6°

If possible, boil or bake meat (a hot dog) and test its internal temperature with a meat thermometer.

Boil together one-half cup of sugar and one cup of water. Test with a candy thermometer.

Use dial- or clock-type thermometers to check the oven and refrigerator temperatures against those given by the built-in temperature controls.

Look for other temperature control devices—for example, heater, air conditioner, freezer.

Use a cooling fan and an electric hair dryer to note the effects of air (wind) on different thermometers.

For practice in converting Fahrenheit to Celsius, see "Think and Do" #5.

AN EASY THERMOMETER

Materials: a clear soft-drink bottle, water, red food coloring, modeling clay, clear plastic straw, thermometer, pencil, strip of tagboard.

Activity: Fill the bottle to the top with colored water. Seal the top with clay. Gently push the straw through the

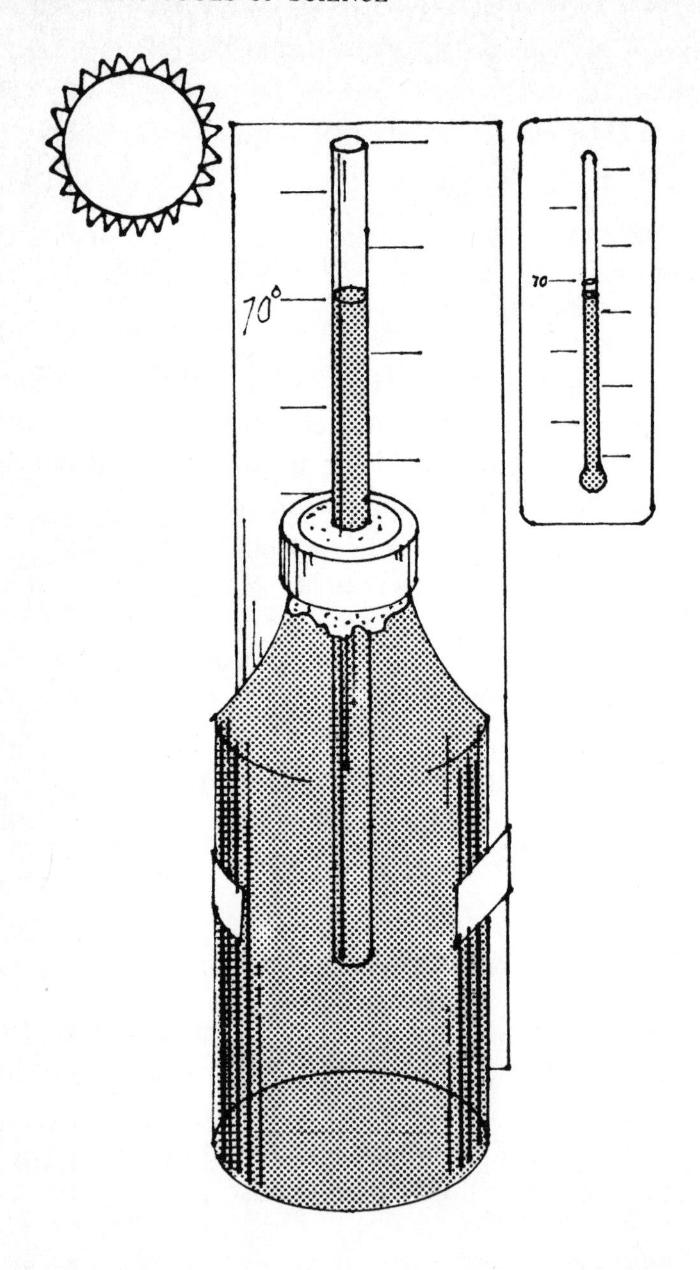
70°
70

clay. Blow through the straw to clear it of any clay. Place the bottle in the sun. As the water heats, it will expand and move up the straw. Place a thermometer alongside the bottle. Tape the strip of tagboard to the back of the bottle. It should be behind the straw. Mark the degrees to correspond with those on the thermometer.

Now you have your own thermometer. Place the bottle in the shade, in the sun, in a pan of warm water, then in a pan of ice. Record the temperatures.

AN EASY BAROMETER

Materials: mercury and dial (aneroid) barometers, one-quart fruit jar, clear soft-drink bottle with a long neck, water, red food coloring.

Activity: A barometer is a tool that is used to measure air pressure. It aids in forecasting changes in the weather.

Fill the jar about two-thirds full of colored water. Turn the empty bottle upside down and insert the neck of the bottle into the jar of colored water.

Set your barometer outside. As the weather changes, the water will rise and fall in the neck of the bottle. The water level will fluctuate more during dramatic weather changes. However, cool mornings, foggy days, and the sun shining on the jar will also give good results.

If the water level falls rapidly, this quite possibly could indicate a storm. If the water level rises, you can probably forecast fair weather. Compare the accuracy of your homemade barometer with that of a commercial one.

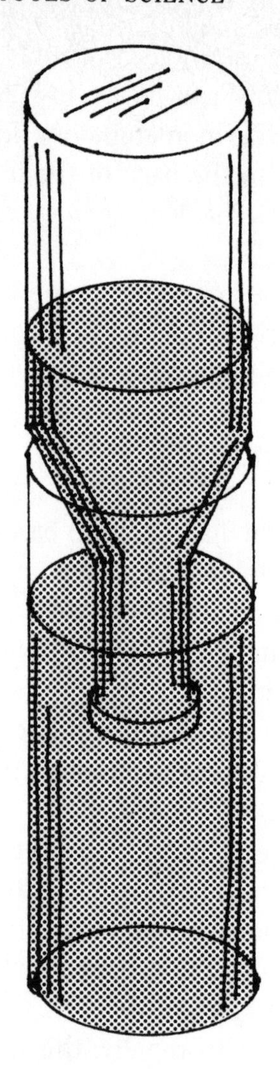

HUMIDITY INDICATOR

Materials: cobalt chloride test papers, cobalt chloride crystals, white paper towels, clear plastic cup, measuring spoon. The test papers can be purchased in 9″ × 12″ sheets and the crystals by the ounce. Both are available at science supply stores.

Activities:

1. Set a white test paper sheet by an open window or outside in the shade. If the air is dry, the paper will turn blue. If there is moisture in the air, it will turn pink.
2. Measure one-quarter teaspoon of cobalt chloride crystals onto a paper towel. Set the towel on the sill of an open window. If the air is dry, the crystals will melt and turn the towel blue.
3. Dissolve one teaspoon of cobalt chloride crystals in one-half cup of water. Dip a test paper strip into the solution. Place the wet strip on the sill of an open window. If the air is dry, the paper will turn blue; if the air is damp, it will remain pink.
4. Pour a spoonful of the solution onto a paper towel and set the towel on the sill of an open window. The reaction will be the same as above—extreme dryness or humidity will cause the colors to deepen.

For added interest, provide several commercial humidity indicators for comparison, observation, and discussion.

CHAPTER 9

MAGNETIC FORCE AS A TOOL

Children love magnets. A magnet is a special form of iron or steel that attracts other pieces of iron or steel. The ends of a magnet are called poles. One end is the north-seeking (N) pole, and the other end is the south-seeking (S) pole. Magnets come in various shapes and sizes. Bar, horseshoe, cylinder, and round magnets are the most common. Avoid very small ones, as they are easily lost or misplaced. Some magnets will lift ten pounds or more.

It is always best to remove your wrist watch before working with a magnet or a compass as the metal in your watch may interfere with your experiment.

MAGNETS AND HOW THEY WORK

Materials: a wide assortment of magnets; iron filings; pieces of iron, steel, aluminum, copper, tin, and other available metals; assorted nonmetal materials—paper, cardboard, wood, plastic, and cork. Other suggested materials: marbles, bones, rubber bands, yarn, bottle caps, toothpicks, keys, buttons, and different kinds and sizes of nails, bolts, and screws. Iron filings can be purchased, collected after rubbing a magnet in most soils, or by pulling steel wool (*not* soap pads) into bits.

Activities: The best way to find out about magnets and how they work is through experimentation. Provide the material and allow children time to investigate and make discoveries.

After children have had an opportunity to play with the magnets, help them to discover the poles. You could determine the N and S by using an already marked magnet. Experiment to find out which two ends repel (push away from each other) and which two ends will attract (pull together). What did you find out about each pole?

THE MAGIC STONE

Materials: a loadstone (lodestone), an assortment of metal and nonmetal material, several different magnets.

A loadstone is a natural magnet that has an N and an S pole and can magnetize some metals by rubbing against them. The ancient Chinese called it "the stone which licks up iron." When large amounts of this stone were found near the Aegean Sea in the region of Magnesia, the people living there called it "the stone of Magnesia," or a "magnet-stone." Find the stone's N and S poles and then see what discoveries you can make with this magic stone.

LET'S MAKE A MAGNET

Materials: large steel needles (darning or sewing machine type), paper clips, steel or iron wire and nails.

Activities: Straighten the paper clips and cut the wire into six-inch pieces. You can magnetize these three items by stroking each with a magnet. Stroke twenty to thirty times in one direction only. Test your temporary magnets to see if they will pick up a paper clip. Does each have an N and an S pole?

Magnetize a piece of wire by stroking. Cut it into four pieces. Can you pick up iron filings with each piece?

WHAT YOU CAN DO WITH A MAGNET

Materials: magnets, iron filings, foam meat trays, plastic wrap, different thicknesses of cardboard, fabrics, carpet pieces, wood, glass, plastic items, leaves, book, paper cup, clear plastic cup, string, paper plate, nails, paper clips, tape, water.

Activities:

1. Test different magnets using foam trays, cardboard, fabric, wood, and so on to see what kind of materials they will pull through.
2. Sprinkle iron filings into a paper plate. Cover the plate with clear plastic wrap and tape the edges securely. Move a magnet about underneath the plate. If the plate is deep enough, you can add other small (metal) objects.
3. Select two different magnets for comparison. Test to see how

many paper clips each will pick up by forming a paper-clip chain. How heavy an object will each pick up? If possible, weigh the objects.

4. Place a paper clip in the plastic cup. Touch a magnet to the outside and move it about to "walk" the paper clip up and down the sides of the cup. Fill the cup with water and find out if the experiment still works.

WHAT DO YOU THINK?

Materials: an assortment of magnets, small bar magnet broken in half, clear plastic cup, piece of thread, small miscellaneous metal items for attracting and repelling. See "Think and Do" #6.

1. Examine a round magnet. Does it have poles? How can you find out?
2. Test to see what part of a magnet is the strongest.
3. Does size have anything to do with a magnet's pull? Test three different sizes.
4. If a bar magnet is broken in half, will each half have an N and an S pole?
5. Place a magnet in a clear plastic cup. Move the cup over different objects. Will it pick up metal objects?
6. Suspend a bar magnet so that it hangs freely. Experiment with both ends of a second magnet. See from what distance the second magnet will pull the first one.

DANCING FIGURES

Materials: shoe box lid or piece of flat cardboard, large paper cup, paper clips, tagboard or construction paper.

Activities: Cut out figures of people or animals about one inch high. Be sure to leave a base at the bottom of each figure. Bend the bottom edge up and attach a paper clip to the base so that each figure will stand. Place figures in the shoe box lid. Move a magnet underneath to make the figures dance. For a round stage, use the bottom of an inverted paper cup.

A small metal washer can also be glued to the base of the paper figure or to small plastic people or animals.

COMPASS MAGNETS

Materials: compasses (hand, pocket, or the type that set on automobile dashboards), darning or sewing machine needle, cork or piece of styrofoam, thread, candle, matches, bowl of water.

Activities: A compass needle is a magnet. The needle points north and south toward the Earth's magnetic poles.

1. Experiment with the compass. Compare it with a magnet. How is it similar? How is it different? Note which end points north.
2. Cut a slice off of one side of the cork so that it will float evenly. Cut a groove in the top side. Magnetize a needle by stroking, then place it in the groove. (Both ends of the needle should extend beyond the cork.) Set the cork in the bowl of water. In which direction does the needle point? Turn the bowl in both directions and note what happens. A flat piece of styrofoam can be used in place of the cork.
3. Attach a piece of thread to the center of a magnetized needle using a drop of hot candle wax. An adult should be present when the candle is lit and used. Suspend the needle where the air is still. In which direction does the needle point?
4. Suspend a bar magnet from its center and hang it directly over a small compass. What happens to the magnet? What happens to the compass?

CHAPTER 10

ELECTRICITY AS A TOOL

A young Tom Edison at your house will particularly enjoy these experiments.

STATIC ELECTRICITY

Static electricity consists of electrons (particles of an atom that carry a negative charge) and ions (atoms that are electrically charged) that do not move.

Materials: plastic comb, bits of tissue paper, glass rod, two balloons, pieces of silk, wool, and fur.

Activities: You can generate your own static electricity in each of the following ways:

1. Walk across a carpet, dragging the soles of your leather shoes. Then touch a metal object. This works best when the air is very dry outside.

2. On a dry day, comb your hair rapidly with a plastic or hard rubber comb.
3. Rub a glass rod with a piece of silk cloth and hold it close to the hairs on your arm.
4. Rub a plastic comb with a piece of wool or fur. The comb will then be able to pick up bits of tissue paper. You can also "charge" a comb by rubbing it rapidly across the fur of a dog or a cat. Be sure not to annoy the animal.
5. Blow up a balloon and tie the end. Rub it across wool, fur, or your hair. If you gave it a "charge of electricity," it will cling to the door or to your outstretched, bare arm. Will your "charged" balloon pick up bits of paper? Cardboard? Plastic?
6. Blow up two balloons. Tie the ends and attach a piece of string to each. "Charge" one of the balloons and then suspend them about a foot apart. (The "charged" balloon is surrounded by a positively charged electric field.) What do you think will happen if you "charge" both balloons? Test your hypothesis.

In the above experiments, as the electrons are rubbed off your hair (wool, fur, etc.), they collect on your comb. The "charged" comb will attract bits of paper. Any hard rubber object can be substituted for the comb.

ELECTROSTATIC ACTION

Materials: plastic cup, nickel, burnt match, comb.

Activity: Stand the nickel on end, carefully place the burnt match across the top edge of the coin, and cover both with the cup. Can you make the match fall from the coin without touching the cup? (Comb your hair rapidly to form an electrostatic charge. Gently rub the side of the cup with the comb. The match will fall off, but the nickel should stay upright.)

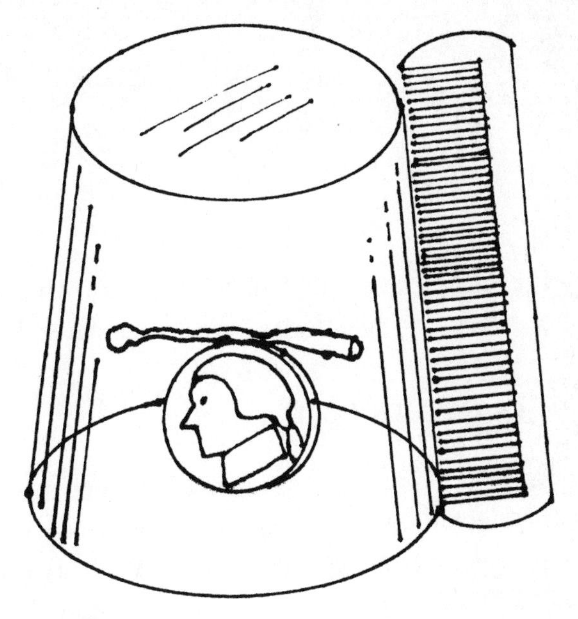

DANCING SNOW

Materials: ten small bits of styrofoam packing, a zip-type plastic bag, piece of wool or fur.

Activities: Place the foam pieces in the bag. Catching the air inside, zip the bag shut. Rub the surface of the bag with wool or fur. Hold the bag in your left hand. Bring one finger of your right hand slowly toward a piece of foam. (The foam will jump away.) Push your finger toward the other foam pieces using a jabbing motion.

While rubbing, your fingers and the foam pieces both developed a negative charge. Two negative charges (or two like charges) repel each other.

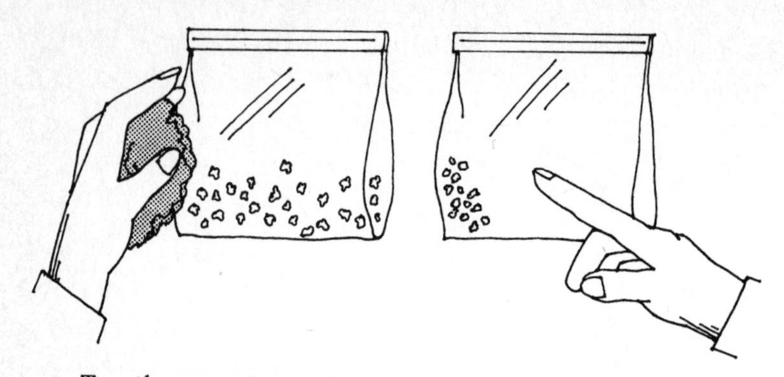

Try the same experiment using the hand that was not used for rubbing the plastic bag. What happens? Why?

NO PLUG-IN ELECTRICITY

Current electricity is made up of *moving* electrons or ions. Most of the electricity we use is current electricity, such as that which powers lamps and appliances. A battery is another source of electric energy.

Materials: two twelve-inch pieces of insulated copper wire, pair of wire strippers, 6-volt dry cell battery, small light bulb and socket with wires attached, collection of metal and nonmetal objects.

Activity: Strip the insulation from each end of the two pieces of wire. Connect one end of each piece of wire to the posts on the dry cell. Touch the other ends to the bare wires extending from the socket. Does the bulb light? Will it light if only one wire is used?

The loop formed by the interaction of the battery, bulb, and wire is called an electric circuit. "Circuit" means going around or a circle. When one wire is disconnected, the circuit is broken, and there is no energy transfer from battery to bulb. This is called an open circuit. A circuit in which there is energy transfer—a bulb lights—is called a closed circuit.

ELECTROMAGNET

Materials: 6-volt dry cell battery, two-inch-long iron bolt, four yards of copper wire, pair of wire strippers.

Activity: If the wire is insulated, strip off all insulation. Leaving a twelve-inch-long tail on the wire, start winding it around the body of the bolt. Continue winding, layer upon layer, using all but the last twelve inches. Twist the two ends to secure the coils to the bolt. Connect one tail wire to each of the posts on the dry cell. Test the head of the bolt to see if you have made a magnet. Test to see how heavy an object it will pick up. Disconnect one wire and see what happens.

ELECTRIC GAME BOARD

Materials: 6-volt dry cell battery, thirteen feet of insulated, copper wire (#18), pair of wire strippers, flashlight bulb, socket with wires attached (a Christmas tree bulb socket with wires attached and a 6-volt bulb works well), two metal-tipped probes, a 14″ × 18″ piece of pegboard, twenty brass paper fasteners with large heads, two 5″ × 16″ strips of paper, tape.

Activity: Cut two fifteen-inch pieces and one eight-inch piece of wire and strip the insulation from the ends. Attach one fifteen-inch piece to the positive (+) post on the battery. Attach the eight-inch piece to the negative (−) post and the other end to the wire on one side of the bulb and socket. Attach the other 15-inch piece of wire to the other wire of the bulb and socket. Cap the ends of both wires with metal probes.

Write ten questions on one strip of paper. Number each. On the second strip of paper, write the answers. Be sure that the answers are not in the same sequence as the questions. Tape the questions to the left side of the board and the answer strip to the right side.

Cut ten wires, each ten inches long. Strip the insulation from the ends. Turn the board over and attach a wire to each fastener in the first row. Attach the other ends of the wires to the matching answers on the second row of fasteners.

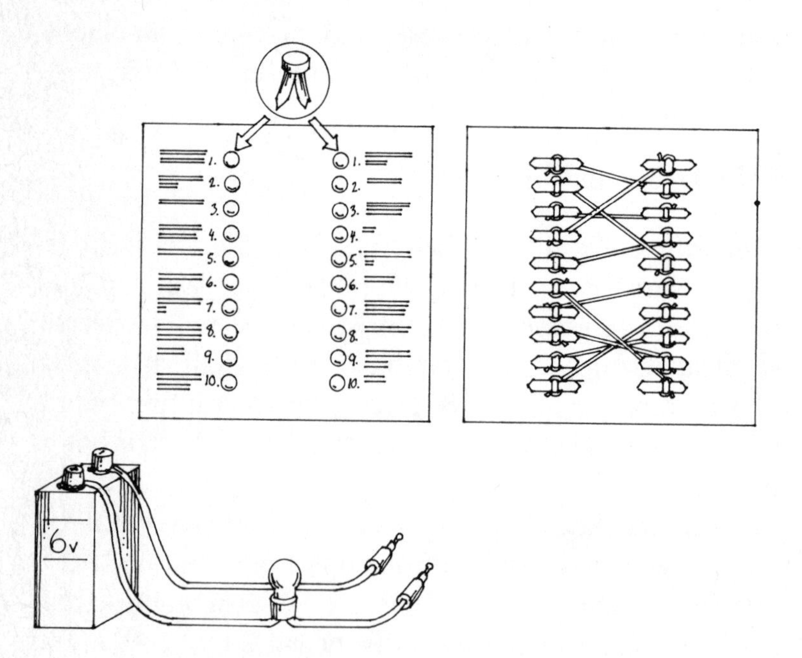

Turn the board to the front side and elevate the top approximately three inches.

Touch one probe to the question and the other probe to the answer. If the answer is correct, the bulb will light.

Question and answer sheets are easily replaced. Rewire the board periodically, as the children will soon memorize the position of the matching wires.

CHAPTER 11

KITCHEN CHEMISTRY AS A TOOL

Did you have fun with popcorn and potato math in Chapter 5? There are lots of science possibilities in your kitchen. It can become your "lab." Working with kitchen utensils and food will give children opportunities to discover many of the physical and chemical changes that take place in the foods they eat and in the way they are prepared.

KITCHEN TOOLS AND HOW TO USE THEM

Stress the importance of carefully following directions, using correct measuring utensils, and making level measurements. Materials should be provided and time allowed to practice measuring skills before having children attempt to measure ingredients for a specific recipe.

Materials: liquid measuring cups in assorted sizes, nested (dry) measuring cups, measuring spoons, plastic knives, two plastic dish tubs, flour, water. See "Think and Do" #7.

Fill one tub with flour and one with water.

Activities:

Liquid Measuring:

1. Use the liquid measuring cups and water to practice measuring. Be exact.
2. Measure to find out how many teaspoons equal a tablespoon.
3. How many tablespoons does it take to equal one-quarter of a cup?

Dry Measuring:

4. Use the nested cups and measuring spoons to practice measurements with flour. Level off by sliding the knife across the top of the full cup or spoon.
5. Suppose a recipe calls for one-quarter cup of flour, but you want to cut the recipe in half. Find out how many tablespoons of flour you will need.

RECIPES

After children have had sufficient practice with the measuring tools, use the flour and water for making paste and clay dough.

Materials: flour, water, salt, plastic containers, scraps of paper, food coloring, wax paper, heavy plastic spoons, measuring cups and spoons.

Activities:

Paste:

Measure two tablespoons of flour into the container.

Add one teaspoon of water. Mix together. (Additional water may be needed.) Tear several small pieces of paper and paste them onto a larger piece, creating your own design.

Clay dough:

Measure one-half cup of flour and one-quarter cup of salt into a container. Mix together. Make a hole in the center of the dry ingredients. Into the hole, pour 2 tablespoons of water and a few drops of food coloring. Mix together. Empty your clay dough onto a sheet of wax paper sprinkled with flour. Knead and form into objects of your choice.

DISPLACEMENT ACTIVITY (HONEY BALLS)

When measuring shortening and peanut butter, try the following method. It is more accurate and makes cleaning the measuring cup easier.

Materials: measuring cup, peanut butter, mixing spoon, powdered milk, honey.

Activities:

1. To measure one-quarter cup of peanut butter, fill a measuring cup three-quarters full of water. Spoon in the peanut butter until the water level reaches a full cup. Hold a spoon over the peanut butter and pour off the water. You will have one-quarter cup of peanut butter and no messy cup. Discuss the displacement process.
2. To make honey balls, place one-quarter cup of peanut butter, one-quarter cup of honey, and one-third cup of powdered milk into a margarine container. Mix well. If the mixture is too sticky, add additional powdered milk. Form into small balls. Eat and enjoy.

EXPERIMENTS IN VOLUME

Cooking spinach and rice and beating egg whites will provide interesting examples that deal with both volume and physical changes.

Materials: hot plate or stove, two cups fresh spinach, one-half cup rice, two large measuring cups, two cooking pots with lids, water, egg, funnel, bowl, rotary egg beater.

Activities:

1. Into one pan, place two cups of spinach and one-quarter cup of water. Cover and cook over low heat for about ten minutes. Remove from heat and pour the contents into a measuring cup. What did you discover?
2. Into the other pan, place one-half cup rice and one cup of water. Stir, cover, and cook for twenty minutes (or follow directions on the box). Remove from heat and pour the contents into a measuring cup. What did you discover?
3. Place the funnel in a measuring cup. Break the egg into the funnel. Set the funnel and egg yolk aside. Note the amount of egg white and then pour it into a bowl. Beat with a rotary egg beater until stiff peaks are formed. What did you discover?

THE GREAT TASTE TEST (YOUR FICKLE TONGUE)

The top of the tongue is covered with tiny bumps. The surface underneath is smooth. Have children feel the difference. The tiny bumps contain the taste buds. There are four types—sweet, sour, salty, and bitter. Different areas of the tongue are associated with different tastes. Study and discuss the drawing of the tongue and then experiment with the following taste tests.

Materials: small paper plates, knife, potato, apple, cocoa, lemon cut into small pieces, salt, sugar.

Activities:

1. Experiment by placing small amounts of salt, lemon juice, sugar, and cocoa on different areas of your tongue. Test one food at a time. Try to determine if the locations of your specific taste buds are similar to those in the drawing.

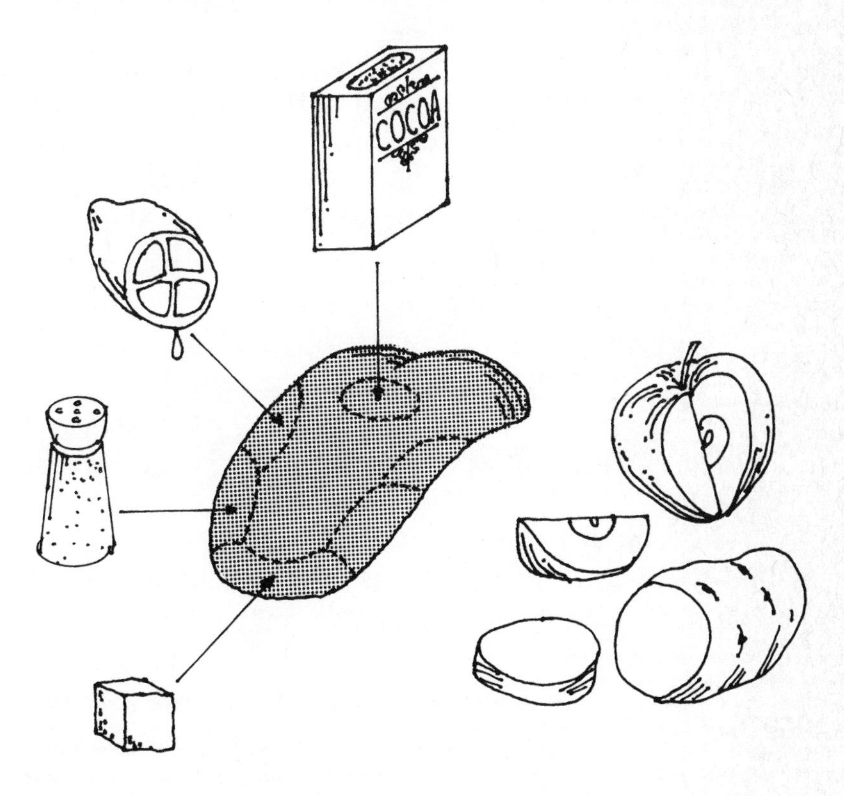

2. Wipe your tongue with a paper towel before re-tasting each of the food items in step 1. Can you taste as well as before? (The tongue is covered with mucus membrane, which mixes with the food to aid us in tasting.)

3. Peel the apple and potato and cut them into same-size cubes. Ask a friend to close her eyes and hold her nose. Place a potato cube and an apple cube on the plate and offer it to your friend. Have her eat one and identify it by the taste. Because taste and smell are closely related, it is often difficult to recognize a food by taste when we cannot smell it.

OXIDATION

Some fruits and vegetables turn brown when they are peeled or cut and left at room temperature. The pigment (coloring) in the cells of the food reacts with the oxygen in the air to cause the discoloring. This process is called oxidation.

Materials: potato, apple, banana, lemon, four vitamin C (ascorbic acid) tablets, knife, two clear plastic cups, three paper plates, water.

Activities: Number the paper plates 1, 2, and 3. Peel and cut three slices each from the potato, apple, and banana. Lay them on plate #1. Dissolve the vitamin C tablets in a cup half filled with water to make an ascorbic acid solution. Squeeze lemon juice into the other glass.

Dip one slice of each of the potato, apple, and banana into the lemon juice and set them on plate #2. Dip one slice of each into the ascorbic acid solution and set them on plate #3.

Wait about fifteen minutes and then compare the three plates. What did you discover? Did plate #2 and plate #3 have the same results? Why? (Lemon juice also contains vitamin C or ascorbic acid.) The children may eat the foods on all three of your test plates.

WHAT IS BUTTER?

Let's find out what butter is.

Materials: one-half cup each of whipping cream and skim milk, two half-pint jars with tight-fitting lids, two funnels, two clear plastic cups, knife, crackers.

Activity: Pour the whipping cream into one jar and the skim milk into the other. Have children work in pairs and take turns shaking the jars and noting any changes.

Butterfat droplets will soon start to form on the sides of the jar containing the whipping cream. As you keep shaking, you will observe the droplets being forced together and forming a solid ball. Place a funnel in each glass and pour the contents from each jar into the funnels. The butter will remain in the funnel and the milk will run through. Compare the results. The whipping cream contains a high percentage of butterfat; the skim milk doesn't. You can check this by reading the label on the cartons. Spread your butter on a cracker and have a taste. Taste the liquid (sweet buttermilk) in the butter glass.

DISSOLVE OR NOT DISSOLVE: THAT'S THE TEST

Some kitchen materials dissolve readily in water; others will not. Test the materials listed below. See "Think and Do" #8.

Materials: clear plastic cups, stirring spoons, teaspoons, cocoa, salt, baking soda, cinnamon, cornstarch, flour, sugar, pepper, powdered sugar, water.

Activities: Predict which materials will dissolve in water. Record your predictions.

Pour one-quarter cup of water into each cup. Measure one teaspoon of material into the water. Use a different spoon to stir and mix each material. Record your observations.

Add an additional one-third cup of cornstarch to the cup containing cornstarch and water. Stir. Keep adding cornstarch, one teaspoon at a time, and stirring until you have a thick mixture. What is happening to the mixture?

Experiment by placing some of the mixture in your hand and allowing it to run through your fingers. Squeeze it and then open your hand. Form it into a ball. Roll it between your palms. Enjoy it.

This unusual mixture flows easily but cannot be poured like a liquid. It is called a "shear thickening fluid." The sudden shearing of the fluid, increasing the viscosity (resistance to flow), is what makes this mixture so fascinating.

ACID/BASE EXPERIMENTS

You can determine if a solution is acid or base by using litmus test papers. Litmus is a dye made from tiny plants called lichens. A base will turn red litmus paper blue. An acid will turn blue litmus paper red. If a solution does not change either of the papers, it is said to be neutral.

Materials: blue and red litmus test papers, vinegar, assorted fruit juices, baking soda, sugar, baking powder, salt, six clear plastic cups, stirring spoons, teaspoons, eyedroppers, water.

Activities:

1. Pour small amounts of fruit juice and vinegar into separate cups. Dissolve one teaspoon each of baking soda, sugar, baking powder, and salt in cups one-quarter full of water. Place a drop from each glass first on the blue test paper and then on the red test paper. (Use a separate dropper for each liquid.) Observe. Did you make any neutral solutions?
2. Set the cup containing the baking soda solution on a newspaper. Add another teaspoon of soda and mix well. Pour the vinegar into the soda solution. What physical reaction do you observe? What chemical reactions? (This chemical reaction forms a gas, and the bubbles rise to the surface.) Add another teaspoon of soda to the glass. Test the solution with litmus papers. Can you explain the test results?

CARBON DIOXIDE IN YOUR BATTER

Leavening is a substance used in cakes, cookies, and bread to make them rise. A gas, carbon dioxide, is formed and bubbles through the batter while it is baking. Baking soda and baking powder are common leavenings used in doughs and batters.

Materials: baking powder, vinegar, baking soda, flour, water, teaspoons, mixing spoons, measuring cup, clear plastic cups, muffin tin, oven.

Activity: Measure one-quarter cup of water into each of three containers. Number each container.

To #1, stir in one teaspoon of baking powder.

To #2, stir in one teaspoon of vinegar and one teaspoon of baking soda.

To #3, add nothing.

Add one-quarter cup of flour to each container. Mix well. Compare the three mixtures. To further test the action of the leavenings, pour the batter into muffin tins and bake at 350 degrees for about fifteen minutes. Have the children predict what will happen to each.

For an edible experiment with leavening, make the Happy Face Cake described in the following section.

HAPPY FACE CAKE

Materials: flour, cocoa, soda, sugar, salt, cooking oil, vinegar, vanilla, nine-inch-square baking pan, measuring cups and spoons, mixing spoon, water, oven.

Activity: To add interest, promote unity, and stress dependency when working together on a project, have children work in pairs. Assign

Pair #1 to measure.

Pair #2 to mix and beat.

Pair #3 to bake and serve.

Pair #4 to assemble materials and preheat the oven to 350 degrees.

Ask another person to read the recipe aloud. That person should read slowly and allow time for the directions to be carried out.

Recipe: Into a nine-inch-square pan, measure one-and-one-half cups of flour, three tablespoons of cocoa, one teaspoon of soda, one cup of sugar, and one-half teaspoon of salt. Mix together.

Make three holes in the dry mixture (a happy face).

Into hole #1, measure five tablespoons of oil.

Into hole #2, measure one tablespoon of vinegar.

Into hole #3, measure one tablespoon of vanilla.

Pour one cup of cold water over the top.

Beat with a spoon until nearly smooth and you cannot see the flour.

Bake at 350 degrees for about thirty minutes. Cool, cut, and serve.

THE MAGIC OF GELATIN

Experimenting with gelatin will provide some interesting observations on one of America's favorite desserts.

Materials: three packages of plain gelatin, a three-ounce package of flavored gelatin dessert, measuring cups and spoons, nine-inch-square baking pan, hot and cold water.

Ask the children what they think gelatin is and where they think it comes from. Examine powdered gelatin. (It is hard, tasteless, and odorless. Gelatin has a high protein content and is made from the bones and hooves of animals.)

Dissolve one tablespoon of dry gelatin in one-quarter cup of cold water. Dissolve one tablespoon of dry gelatin in one-quarter cup of hot water. Let each set about five minutes and then compare.

Wiggle Jell: Empty two packages of gelatin into one-quarter cup of cold water. Stir until dissolved. Let it set for five minutes. You should be able to safely turn the cup upside down.

Boil one cup of water. Pour in the three-ounce package of flavored gelatin. Stir until dissolved. Add the plain gelatin to this mixture. Mix well. Pour into a 9″ × 9″ cake pan. Place in the refrigerator for about fifteen minutes. Cut into squares. Pick it up in your hand. Note the changes before you eat it.

Children can experiment by adding the gelatin-and-cold-water mixture to fruit juices or other liquids.

PART THREE

A TOOL CATALOG

CHAPTER 12

PUZZLER PAGES

Here are some rainy-day learning games for young scientists:

OCCUPATIONS

Match the occupations with the basic tool each needs.

astronomer	binoculars
scientist	barometer
physician	scale
photographer	telescope
meteorologist	stethoscope
birdwatcher	periscope
submariner	camera
metrologist	microscope

OPPOSITES

Write the opposite for each word in the blank spaces below. Read each row of letters from bottom to top until you find a certain kind of thermometer.

1. north — — — — —
2. down — —
3. low — — — —
4. minus — — — —
5. fast — — — —
6. attract — — — — —
7. hot — — — —

SCOOP ON SCOPE

Write in the correct "scopes" to fit the blank spaces. If you need help, check chapters 4 and 7.

1. m — — — — scope
2. — e — — scope
3. — — — t h — scope
4. — — — e scope
5. — a — — — scope

LARGEST AND SMALLEST

The girl wearing the blue dress has a magnifying glass larger than the girl in the striped dress.

The girl in the polka-dot dress has a smaller magnifying

glass than the girl in the plaid dress but a larger one than the girl in the striped dress.

The girl in the plaid dress has a magnifying glass larger than the girl in the polka-dot dress.

The girl in the striped dress has a magnifying glass smaller than the girl in the plaid dress.

Which girl has the largest magnifying glass?

Which girl has the smallest magnifying glass?

SCRAMBLED SCIENCE

Unscramble each word below to find the science-related word.

1. namget
2. laceban
3. micetrysh
4. ursmeea
5. teerm
6. racema
7. wainrob
8. materbroe

WEATHER BOGGLE

Find these weather terms in the puzzle below: *rain, hail, wind, snow, sun, cold, fog.* Spell out each word, drawing a continuous line from letter to letter. You may go up, down, left, right, and on a diagonal—but you may not jump over any letter. Letters can be used more than once.

H I W O

A L N S

O R D U

C G O F

CLOCK TALK

Write the correct answers in the blank spaces below. Read the letters in the vertical block, from bottom to top, to find a type of clock.

1. tells time — _ [_] _ _ _
2. good to drink — _ [_] _ _ _
3. clock face — _ [_] _ _
4. follows me — _ _ _ [_] _ _
5. direction — [_] _ _ _ _
6. energy source — _ [_] _
7. experiment — _ _ [_] _

EXTRA, EXTRA

Write the extra letter in the space beside each word. Read the letters from top to bottom to find an extra word.

1. barommeter __
2. diail __
3. therrmometer __
4. norrth __
5. microscoope __
6. batterry __

TOOL ASSOCIATION

Match each word to the picture to which it is the most closely associated.

1. zero
2. colors
3. dry cell
4. reflection
5. weight
6. shadow
7. metal
8. needle
9. stars

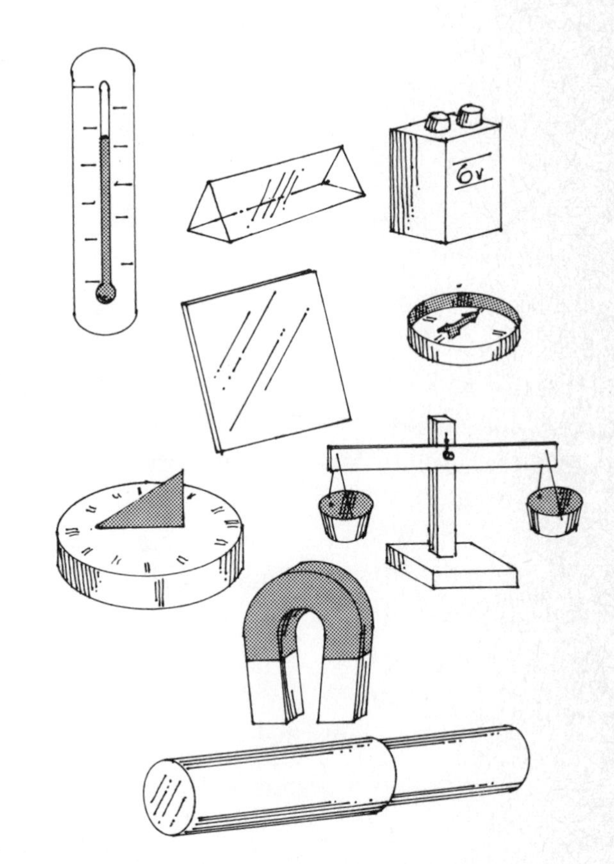

WORDCROSS

Across
1. below zero
2. direction finder
3. solar energy

Down
1. timer
2. clock face
3. magnifying glass

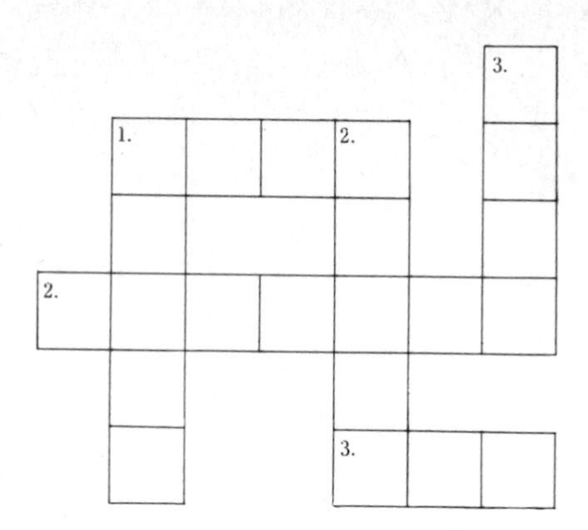

A PEANUT PROBLEM

The boy with red hair had fourteen peanuts and ate half of them.

The boy with the black hair had twice as many peanuts as the boy with the red hair and ate twice as many as were eaten by the boy with the blond hair.

The boy with brown hair had half as many plus 2 as the boy with the black hair and he ate $3 \times 2 + 4$ peanuts.

The boy with the blond hair had more peanuts than the boy with the red hair but fewer than the boy with the brown hair, and he ate one less peanut than was eaten by the boy with the red hair.

Fill in the blanks below.

Boy's hair	# peanuts	# eaten	# left over
Red	______	______	______
Black	______	______	______
Brown	______	______	______
Blond	______	______	______

Which boy had the least number of peanuts? ________
Which boy had the most peanuts left over? ________

MAGNET SEARCH

Find and circle these words in the horseshoe: *bar, iron, magnet, pole, north, pulls, repel, nail.* You can go up, down, right, or left. Letters can be circled more than once.

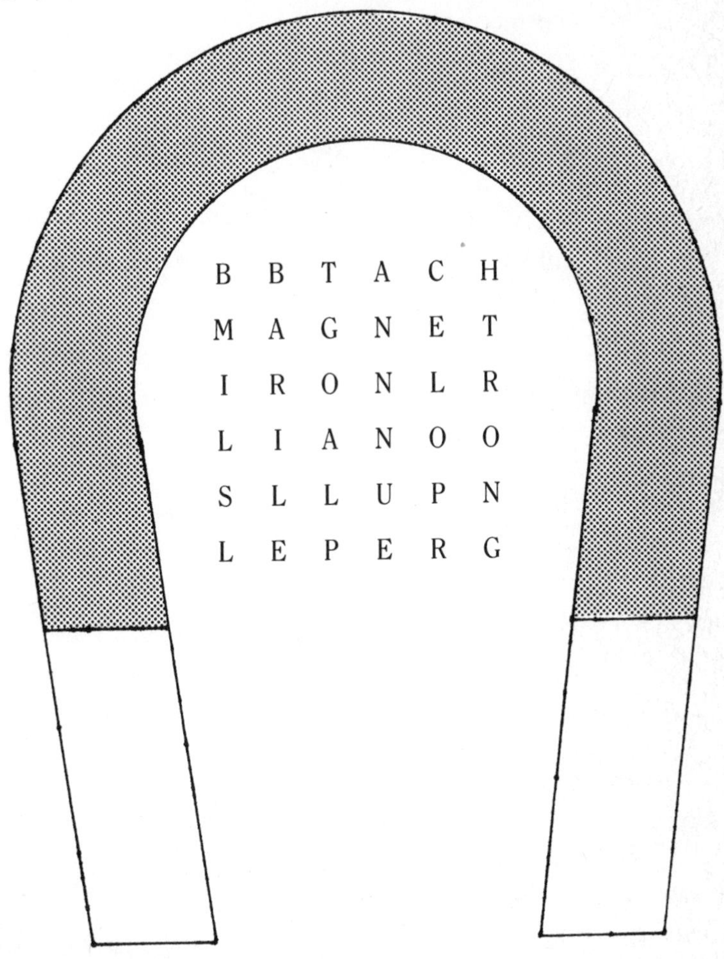

SCIENCE HUNT

Find and circle the words listed below. The words can be found in a line, either forward, backward, up, down, or diagonally. Letters can be circled more than once.

magnet	scale	battery
hot	measure	bar
meter	thermometer	test
compass	telescope	big
lens	cone	cold
inch	camera	green
estimate	feet	prism
predict	arch	down
clock	mirror	up

```
A X X B E R U S A E M
R X B I X M S I R P I
C M A G N E T N U O R
H O T X C L O C K C R
X T T B A R X H L S O
T H E R M O M E T E R
X G R E E N N C O L D
T I Y X R S C A L E O
S L S S A P M O C T W
E S T I M A T E N X N
T C I D E R P X X E R
```

ANSWERS FOR PUZZLER PAGES (CHAPTER 12)

OCCUPATIONS

1. astronomer: telescope
2. scientist: microscope
3. physician: stethoscope
4. photographer: camera
5. meteorologist: barometer
6. birdwatcher: binoculars
7. submariner: periscope
8. metrologist: scale

OPPOSITES

1. north–south
2. down–up
3. low–high
4. minus–plus

thermometer: celsius

5. fast–slow
6. attract–repel
7. cold–hot

SCOOP ON SCOPE

1. microscope
2. periscope
3. stethoscope
4. telescope
5. kaleidoscope

LARGEST AND SMALLEST

The owner of the largest magnifying glass is the girl in the blue dress.

The owner of the smallest magnifying glass is the girl in the striped dress.

SCRAMBLED SCIENCE

1. magnet
2. balance
3. chemistry
4. measure
5. meter
6. camera
7. rainbow
8. barometer

WEATHER BOGGLE

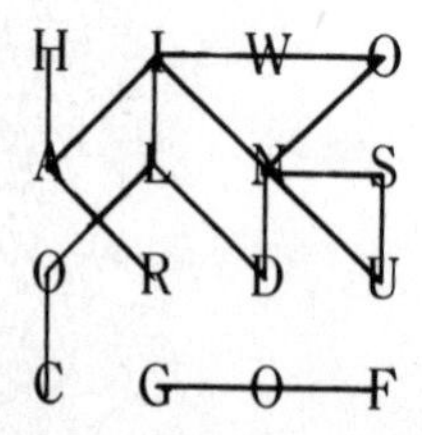

CLOCK TALK

1. clock
2. water
3. dial
4. shadow
5. north
6. sun
7. test

type of clock: sundial

EXTRA, EXTRA

1. barommeter	m
2. diail	i
3. therrmometer	r
4. norrth	r
5. microscoope	o
6. batterry	r

TOOL ASSOCIATION

1. zero: thermometer
2. colors: prism
3. dry cell: battery
4. reflection: mirror
5. weight: balance scale
6. shadow: sun
7. metal: magnet
8. needle: compass
9. stars: telescope

WORDCROSS

						L
	C	O	L	D		E
	L			I		N
C	O	M	P	A	S	S
	C			L		
	K			S	U	N

A PEANUT PROBLEM

Boy's hair	# peanuts	# eaten	# left over
Red	14	7	7
Black	28	12	16
Brown	16	10	6
Blond	15	6	9

The boy with the red hair had the least number of peanuts.
The boy with the black hair had the most peanuts left over.

MAGNET SEARCH

B	B	T	A	C	H
M	A	G	N	E	T
I	R	O	N	L	R
L	I	A	N	O	O
S	L	L	U	P	N
L	E	P	E	R	G

SCIENCE HUNT

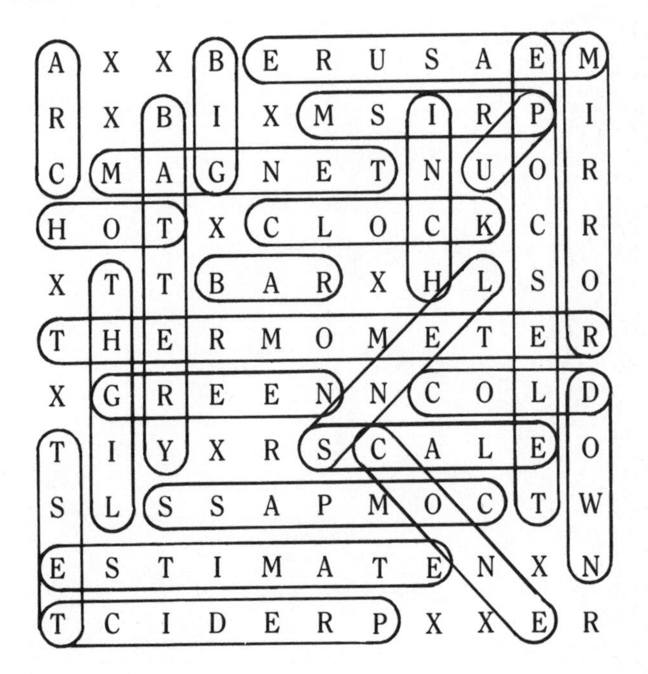

CHAPTER 13

WHIZ QUIZZES AND FASCINATING FACTS

You might try these while on a seemingly interminable automobile trip with your children or any outing with kids that requires activities to pass the time. Good, too, for a young scientist's birthday-party game:

LARGEST AND SMALLEST

1. What is the largest living animal?
2. What is the smallest mammal?
3. What is the largest living land animal?
4. What is the tallest living animal?
5. What is the largest North American bird?
6. What is the smallest North American bird?
7. What animal has the largest eyes?

8. What animal has the largest ears?
9. What is the largest insect?
10. What is the longest insect?
11. What is the largest shellfish?
12. What is the largest plant?
13. What is the largest seed?
14. What is the smallest seed?

SPEED TEST

1. What is the fastest animal on four legs?
2. What is the fastest fish?
3. What is the fastest insect?
4. What is the fastest bird?
5. What is the fastest thing in the universe?

NUMBERS GAME

Our body is a fascinating machine. How well do you know what's going on inside of you? Choose the correct answer for each of the following questions.

1. What percent of your body is water?
   65% 50% 45%
2. What is the number of tones our ears can discriminate among?
   20,000 100,000 300,000
3. What is the number of differences in color our eyes can distinguish?
   4 million 6 million 8 million
4. What is the number of teeth in a complete adult set?
   28 30 32

5. What is the number of bones in your body?
   156 206 306
6. What is the number of bones in your hand?
   20 27 33
7. What is the approximate number of quarts of blood in an eighty-pound child?
   2½ 4 6
8. Who has the largest number of neck bones?
   you a mouse a giraffe

BRAIN TESTERS

1. What animal can crawl upside-down, builds and carries its own house, and has more than 30,000 teeth?
2. What creature has no head, eyes, or teeth and lives up to 150 years?
3. Does water swirl clockwise or in a counterclockwise direction as it goes down the drain?
4. Who is the strongest—you, an ant, a bee, or a snail?
5. Are crickets right- or left-winged?
6. Which is the best jumper—a kangaroo, a frog, a flea, or a grasshopper?

FASCINATING FACTS

FUNNY FACE. Your face is capable of making 250,000 different expressions. Look in the mirror and see how many different ones you can make.

ANIMAL FACTS. A polar bear is as tiny as a guinea pig at birth. But when fully grown, it may weigh as much as half a ton and stand as tall as eight feet on its hind legs.

A sea otter can swim on its back while balancing a rock on its chest. This crafty little fellow will then take a shellfish in its paws and pound it against the rock. The otter eats the soft meat and tosses away the shell.

Some varieties of aphids are called ant cows, but ants don't eat aphids. They milk them for their honeydew.

HEAVY HEAVY. The earth weighs 6,570,000,000,000,000,000,000 tons. Can you read this figure? (6 sextillion, 570 quintillion)

COOL IT. If a beehive gets hot, the cells that hold honey will melt and the honey will be lost. To prevent this, squads of bees take turns beating their wings, 11,400 times a minute, at the entrance to the hive. This reduces the temperature and produces a humming sound.

The African elephant's ears weigh more than 100 pounds each. As they flap, blood running through their countless blood vessels is cooled by air, and the elephant's body temperature is lowered.

ELECTRONIC EASELS. Art students at Carnegie-Mellon University in Pittsburgh can draw on a computerized canvas with an electronic palette featuring 16.8 million colors.

KING CACTUS. There are more than 1,000 kinds of cactus in the American desert. The king of them all is the saguaro. It grows to 50 feet tall and can hold up to 10 tons of water.

POPPING GOOD. What makes popcorn pop? Popcorn pops because each kernel contains a tiny drop of water. When the grains are heated, this moisture turns to steam and results in a mini-explosion that splits the kernels open.

MAGNETS, MAGNETS EVERYWHERE

Magnets in a cow? Yes. Most dairies insert bullet-shaped magnets in the first section of the stomach of their milk cows. The magnet collects any piece of metal that might be in the hay or grass eaten by the cows. Their strong digestive juices will dissolve the metal, so it doesn't become embedded in the intestines and cause infection or death.

Do honeybees carry magnets? Scientists experimented by first holding bees next to a large magnet. Then when they tested them with a special instrument, they discovered that some bees seem to have a natural built-in magnetism. They are not yet sure if some bees might have one or more tiny magnets or that it might be that they have tiny iron crystals located in special cells.

Scientists have also found magnetic tissue in pigeons, salmon, and dolphins.

Do you have a compass in your nose? Maybe. A British biologist at the University of Manchester discovered deposits of iron in the lining of human noses. This material seems to act like a compass. One theory is that this material may help us detect directions in relation to Earth's magnetic field.

Don't be surprised if, one day soon, when you go to the doctor for a medical examination, the doctor passes a nine-ton magnet over your body. Scientists at General Electric Company have developed a magnet and sound wave system that will allow doctors to see through bones and into cells. This method is far superior to X-rays or scanning devices.

Scientists are also killing cancers in animals with heat

from tiny magnets implanted in tumors and warmed by magnetic fields. Could tiny magnets possibly provide a means for wiping out cancer completely?

ANIMAL OLYMPICS

Maybe you've heard of the famous frog-jumping contest held in May of each year at the Calaveras County Fairgrounds in Angels, California, but do you know about the annual snail, maggot, roach, and turtle races?

Cockroach racing in Biloxi, Mississippi, is a fascinating sport, and often a great deal of money changes hands at these events. Roaches that have been trained to race are worth a lot of money, and the winning roach often ends up embalmed, gold-plated, or mounted on red velvet.

In Prichard, Idaho, each February, the world championship maggot races are held as part of this community's Lumberjack Day celebration. Maggots are placed in the middle of a circle, and the first to reach the perimeter is the winner.

The racing course is the asphalt parking lot outside Brennan's Bar at Marina Del Rey in Los Angeles, California. Each Thursday evening up to 700 turtles gather with their proud owners to take part in thirteen races. The course is a circle sixteen feet in diameter. The winner gets a plywood trophy, and the turtle's time and name are engraved on a plaque hanging behind the bar. Some of these turtles have been racing for five years.

Murillo de Rio Leza, Spain, is the home of the international snail-racing championships. The winner is the snail that is first to reach the end of a 1.6-yard course. The Mountain Climb race is also held here. The prize goes to the snail that is first to reach the top of a ramp.

ANSWERS FOR WHIZ QUIZZES AND FASCINATING FACTS (CHAPTER 13)

LARGEST AND SMALLEST

1. The largest living animal is the blue whale. It can grow as long as the length of three school buses parked end to end (about 100 feet) and weigh as much as 30 elephants (about 200 tons).
2. The smallest mammal is the pygmy shrew. The tiniest of these may weigh as little as a penny.
3. The largest living land animal is the African elephant. An average male is about 10 feet tall and weighs about 6 tons. It may eat as much as 350 pounds of food each day and drink 30 to 50 gallons of water.
4. The tallest living animal is the giraffe. Giraffes often reach a height of 18 feet.
5. The largest North American bird is the California condor. It measures 45 to 55 inches long and has a wingspread of 102 to 114 inches. The condor is also the rarest bird. There are only 20 to 30 surviving.
6. The smallest North American bird is the Calliope hummingbird. It measures 2¾ to 3½ inches long with a wingspread of only 4¼ inches. This fascinating bird can dart at 60 miles per

hour, can hover like a helicopter, and can fly backwards and sideways. The tiny Calliope lays eggs the size of navy beans in a nest no bigger than a teaspoon.

7. The largest eyes in the world belong to the great blue whale. Its eyeballs measure about 5 inches in diameter.
8. The animal with the largest ears is the African elephant whose ears weigh more than 100 pounds each.
9. The largest insect is the African Goliath beetle. It is 4 to 6 inches long and weighs from 2 to 3 ounces.
10. The longest insect is the walking stick of Australia. It grows to 12 inches in length.
11. The largest shellfish is the giant Pacific (tridanca) clam. Some species weigh more than 400 pounds and live as long as 100 years.
12. The largest plant in the world is the giant sequoia. The largest one, named General Sherman, is figured to weigh more than 2,000 tons. It is 272 feet tall and measures 79 feet around its trunk.
13. The largest seed is the coconut. It grows from 8 to 12 inches long and 6 to 10 inches across.
14. The smallest seed is the tiny seed of the epiphytic orchid. It takes 35,000,000 seeds to equal one ounce.

SPEED TEST

1. The cheetah is the fastest animal on four legs. It has been clocked to speeds of 70 miles an hour. When it runs, only one foot at a time touches the ground.
2. The sailfish is considered the fastest fish. It has reached speeds of more than 68 miles an hour.
3. The speediest insect is the dragonfly. Some dart by at more than 50 miles an hour.

4. The golden eagle claims the title of the fastest bird, flying at approximately 120 miles an hour.
5. Light, the fastest thing in the universe, travels at 186,282 miles per second. That's more than 7 times around the Earth.

NUMBERS GAME

1. Humans are about 65 percent water.
2. Our ears can discriminate among 300,000 tones.
3. Your eye can distinguish nearly 8 million differences in color.
4. There are 32 teeth in the complete adult set.
5. The body contains 206 bones. This includes the 3 floating bones in each ear.
6. Each human hand contains 27 bones.
7. There are about 2½ quarts of blood in the body of an 8-year old.
8. Humans, giraffes, and mice all have the same number of neck bones—7.

BRAIN TESTERS

1. A garden snail can crawl upside-down, builds and carries its own house, and has more than 30,000 teeth. It can also live 3 to 4 years without eating or drinking.
2. A clam has no head, eyes, or teeth, and some types can live up to 150 years. It breathes and eats (and the female lays eggs) through its neck.
3. In the Northern hemisphere, water swirls counterclockwise going down the drain. In the Southern hemisphere, it turns

clockwise. If the tub's drain sat directly on the equator, you would find the water twisting in both directions.

4. If you were as strong as an ant and weighed 80 pounds, you could lift 2 tons or one large automobile.

   A bee can carry 300 times its weight. If you weighed 80 pounds and could carry 300 times your weight, that would equal 24,000 pounds.

   A garden snail weighs about an ounce but can pull 200 times its weight and carry 12 times its weight. If you were as strong as a garden snail and weighed 80 pounds, you could pull 16,000 pounds and carry 960 pounds.

   Now can you figure out who is the strongest?
5. Among the 2,400 species of crickets found across the world, 95 percent of the males were "right-winged." The male makes his special music by rubbing his notched right wing over his sharp-edged left wing.
6. Grasshoppers, kangaroos, and frogs are all great jumpers. But for their size, fleas are the best jumpers of them all. A flea can jump over 10 inches. This is about 100 times its height. How far would you be able to jump if you could jump 100 times your height?

CHAPTER 14

THINK AND DO WORKSHEETS

Kids: It's okay to copy these pages and put them into your own "Lab Manual." Then you'll have a personal album of your experiments and the results.

THINK AND DO #1

Using a Stethoscope

Use a stethoscope and a timepiece with a second hand for each activity below. Work with a friend and do the timing for each other. For the first column, sit quietly so that your body is functioning at a normal rate. For the second column, jump up and down ten times and take the readings again. Time each activity for fifteen seconds and then multiply by four to equal one minute. This method will be more accurate than timing for one full minute. Record your answers in the spaces provided below.

Your	Sitting quietly	After exercising
Heartbeat	____ × 4 = ____	____ × 4 = ____
Pulse (inside elbow)	____ × 4 = ____	____ × 4 = ____
Breathing	____ × 4 = ____	____ × 4 = ____
A friend's		
Heartbeat	____ × 4 = ____	____ × 4 = ____
Pulse (inside elbow)	____ × 4 = ____	____ × 4 = ____
Breathing	____ × 4 = ____	____ × 4 = ____

Compare your recording with that of your friend. Compare your recording with those of other members of your class. Your comparisons in the first column should be fairly close. Those in the second column could show a wide variation. Can you explain this?

THINK AND DO #2

Popcorn Math

Think carefully and then write your estimates in the spaces below.

Estimates

Number of kernels landing in your space. ______

Distance of kernels popping the farthest in your space. ______

Number of kernels landing in circle #1. ______

Number of kernels landing in circle #2. ______
Number of kernels landing in circle #3. ______
Total ______

After the corn pops, count and record the actual numbers in the spaces below. Multiply the kernels landing in each circle by their assigned point value. Total your score.

Results

Number of kernels landing in your space. ______
Distance of kernels popping the farthest in your space. ______
Number of kernels landing in circle #1. ______ × 1 = ________
Number of kernels landing in circle #2. ______ × 2 = ________
Number of kernels landing in circle #3. ______ × 5 = ________
Total ________

Compare your estimates with the actual results. Do you think you are pretty good at estimating?

THINK AND DO #3

Peanut Math

Share a jar of peanuts with three other people. Read the instructions and questions carefully and record your answers in the spaces provided.

1. Estimate the number of peanuts in the jar. ______
2. Pour a pile of peanuts onto each paper towel. Estimate the number of peanuts in your pile. ______

3. Count the peanuts in each pile and find the total.
   ______ + ______ + ______ + ______ = ______
4. Was your estimate more or less than the actual number of peanuts in the jar? ______
   By how many peanuts? ______
5. Divide the total number of peanuts by 4. Give each person an equal share.
   How many peanuts do you have? ______
6. Estimate the number of whole peanuts inside your shells. ______
7. Shell your peanuts. How many do you have? ______
8. Was your estimate more or less than the actual number of whole peanuts? ______
   By how many? ______
9. Estimate the weight of your shells. ______
   Estimate the weight of your shelled peanuts. ______
10. Weigh your shells and peanuts separately.
    How much do the shells weigh? ______
    How much do the peanuts weigh? ______
11. Set up a math example based on the ratio of peanuts to shells.

THINK AND DO #4

Potato Math

Place your potato piece in a container and mash until smooth. Discuss your answers with your fellow scientists.

1. Predict what will happen if you add powdered sugar to your mashed potato.
2. Add one level tablespoon of powdered sugar and mix well. What changes do you see taking place?

3. Estimate the number of tablespoons of powdered sugar that you will need to add before this mixture will form a firm, nonsticky ball.
4. Add another tablespoon of powdered sugar and mix well. Continue this process, recording each spoonful added, until you have a firm, nonsticky ball.
   Make tally marks on a piece of paper.
5. How accurate was your prediction?
   Describe your reaction to this experiment.
6. Predict what would happen if you added six tablespoons of each of the following to a cooked mashed-potato cube. Then do the experiments and compare the results with your predictions.
   a. granulated sugar
   b. brown sugar
   c. cornstarch
7. Predict what would happen if you added six tablespoons of each of the following to a cooked, mashed sweet potato cube. Then do the experiments and compare the results with your predictions.
   a. powdered sugar
   b. granulated sugar
   c. brown sugar
   d. cornstarch

THINK AND DO #5

Changing Fahrenheit to Celsius

A Fahrenheit degree is smaller than a Celsius (centigrade) degree. One F degree = 5/9 of a C degree.

To convert C to F, multiply by 9, divide by 5, and add 32.

To convert F to C, subtract 32, multiply by 5, and divide by 9.

Find the answers to the examples below.

1. 59 F − 32 = ______ × 5 = ______ − 9 = ______ degrees C.
2. 40 C × 9 = ______ − 5 = ______ + 32 = ______ degrees F.
3. Jenny looked at the outdoor thermometer. It read 140 degrees C. "That's 65 degrees F," she said. Was she right? If not, what should she have said?
4. "It's 0 degrees C," said Carlos as he pulled the thermometer out of the ice bucket. The thermometer read 32 degrees F. Was Carlos right?
5. "The thermometer reads 100 degrees F. Why isn't the water boiling?" asked Maria. Can you answer her question?
6. Taro pulled the thermometer out of the glass. "Hey, look!" he exclaimed. "This F thermometer reads 0 degrees." What do you think was in the glass?

THINK AND DO #6

What Do You Think?

Examine several different types and sizes of magnets and a bar magnet that has been broken in half.

1. Look closely at several round magnets.
   a. Do you think round magnets have poles?
   b. How can you test to find out?
   c. Test the magnets. Was your answer in (a) right?
2. Test several magnets to find out which part of the magnet has the strongest pull. What did you find out?
3. Does size have anything to do with a magnet's pull? Measure and then test three different sizes of the same type of magnet

(round, bar, or horseshoe). Record your findings. What did you discover?

4. Do you think a bar magnet that has been broken in half will have an N and an S pole in each half? Why? Test the magnet pieces. What did you discover?

THINK AND DO #7

Kitchen Tools

Measuring spoons and cups and pint, quart, and gallon containers will be needed for the following activities.

Measure carefully to find the correct answers. Can you figure some of the answers without measuring?

The number of

1. Teaspoons to equal 1 tablespoon.
2. Tablespoons to equal ¼ cup.
3. Cups to equal 1 pint.
4. Pints to equal 1 quart.
5. Quarts to equal 1 gallon.
6. Tablespoons to equal ½ cup.
7. Cups to equal 1 quart.
8. Tablespoons to equal ¾ cup.
9. Cups to equal ½ pint.
10. Tablespoons to equal 1 cup.
11. Pints to equal 1 gallon.

CHOCOLATE PUDDING

Suppose you want to double this recipe for chocolate pudding and your friend wants to cut it in half. Write down the correct measurements. Be sure to list teaspoons or table-spoons.

4½ teapoons flour
¼ cup sugar
1 tablespoon cocoa
½ pint of milk
1 egg yolk

If you want to make the chocolate pudding, mix the cocoa, flour, and sugar together. Beat the egg yolk with a fork and add it to the milk. Slowly add the liquid to the dry ingredients. Stir until well blended. Cook over medium heat until thick. Cool before you eat. The original recipe serves two.

THINK AND DO #8

Dissolve Or Not Dissolve

Predict which materials will dissolve in water. Then test by adding one teaspoon of each to one-quarter glass of water. Stir well. Record your predictions and test your results.

cocoa
salt
baking soda
cinnamon
flour
sugar
pepper

powdered sugar
cornstarch

Add an additional one-third cup of cornstarch, one teaspoon at a time, to the glass containing cornstarch and water. Stir after each addition. What did you observe?

APPENDIX

Children's science magazines are fun to read and will enhance any science program by providing additional information, current events, beautiful photographs, and explanatory drawings. Short reviews, addresses, and notations on those accepting contributions from children are listed below and denoted with an asterisk.

SCIENCE MAGAZINES

Dolphin Log is a science discovery magazine published by the Costeau Society for children aged seven to fifteen. Its primary emphasis is on motivating children to develop knowledge and interest in the interrelatedness of all living organisms. New discoveries, fascinating facts, experiments and puzzles—all with an ocean water–related theme—are presented in a humorous and easy-to-understand format.

The illustrations and diagrams will provide an additional understanding of the marine and coastal waterway system. Teachers and parents will value the slip-out, correlated reader's activity and resource sheet, which contains clever crafts, art projects, games, and puzzles. Many beautiful photographs enhance this publication.

National Geographic World, a monthly publication for school-age children, is produced by the School Services Division of the National Geographic Society. It contains many beautiful color photographs along with a number of double-page photo features. Geography, other countries, animals, and cultures are some of the topics covered. The back cover of each issue offers several close-up, partial photos, plus clues, to challenge the perceptual skills of young readers. On the "fun page," you will find things to make, games, and science ideas. You will also read on the "Kids Did" page about kids who have had unusual experiences. The "Far-Out Facts" are incredible!

*Children can write in and ask questions. If their questions are chosen, the answers will appear in a later issue.

Odyssey is published monthly by Astro Media Corporation and is for children aged eight through twelve. This fascinating publication offers many photo features and emphasizes astronomy and outer space. Its pages are filled with how-to projects, experiments, fun-to-do activities, and thrilling reports for young spacepeople. The unique center newspaper section brings you the latest news from NASA, a comic strip adventure, book reviews, and profiles.

*Young readers are invited to send in their comments and questions. Poems, projects, pictures, and stories are also solicited, but only on designated topics.

Ranger Rick's Nature Magazine is published by the National Wildlife Federation for children aged six to twelve. You receive the magazine by becoming a member of Ranger Rick's Nature Club. Additional materials and activity guides are available to teachers. Articles and stories are written about natural history, the environment, and people working for the benefit of animals and the environment. "Ranger Rick and His Friends," a regular feature, provides entertainment as well as information. Because children can relate so easily to the characters, they could be challenged to write additional episodes. The "Nature Club News" brings you up to date on the news on ecology, conservation, and recycling. The poems, puzzles, and exceptional photography will become helpful assets to your science program.

*The editors accept children's letters, written to Ranger Rick, on any science area.

Science and Children magazine for members of the National Science Teachers Association and is published for teachers of the elementary grades. It is not a magazine for children but a resource for teachers. The publication offers many how-to science projects and activities, science photo features, and information on current research in science education. Teachers will find the examples of personal experiences, successful ideas, suggestions for planning and developing science programs, and comments by other teachers of significant value.

Scienceland is a softcover science magazine/booklet for use with children through the third grade. Each volume contains eight issues. Unlike other periodicals, the magazine is designed for repeated use year after year. The delightful photographs and short, easy-to-read text make this an excellent choice for younger children and children with special needs. Added features include a read-and-participate page, unusual facts, guessing games, and science experiments. Each issue contains a vocabulary list and a pronunciation key. Back issues (volumes) can also be ordered. Write for a table of contents sheet that includes all volumes. A separate teacher's guide, containing background information, discussion questions, activities, and projects, is also available.

3-2-1 Contact is a science magazine for children aged eight through twelve. It is published by the Children's Television Workshop, the creators of "Sesame Street." Each issue, containing feature articles, games, puzzles, and photo features, is based on a theme. The beautiful centerfold posters are easily removed for display. This colorful and interesting publication is filled with reports, unusual facts, reviews, and previews. A serialized detective/mystery story makes a good choice for reading aloud to children. (They could write their own story ending!)

*Areas open to young contributors are questions or short items from magazines and newspapers for the "Contact Report" pages and drawings based on the editor's special topics.

Dolphin Log
The Cousteau Society Membership Center
930 West 21st Street
Norfolk, VA 23517

National Geographic World
Dept. 01081
17th and M Streets N.W.
Washington, DC 20036

Odyssey
625 E. St. Paul Avenue
P.O. Box 92788
Milwaukee, WI 53202

Ranger Rick's Nature Magazine
National Wildlife Federation
1412 16th Street, N.W.
Washington, DC 20036

Science and Children
The National Science Teachers Association
1742 Connecticut Avenue
Washington, DC 20009

Scienceland
501 Fifth Avenue
New York, NY 10017

3-2-1 Contact
E=MC Square
P.O. Box 3932
Boulder, CO 80321

INDEX

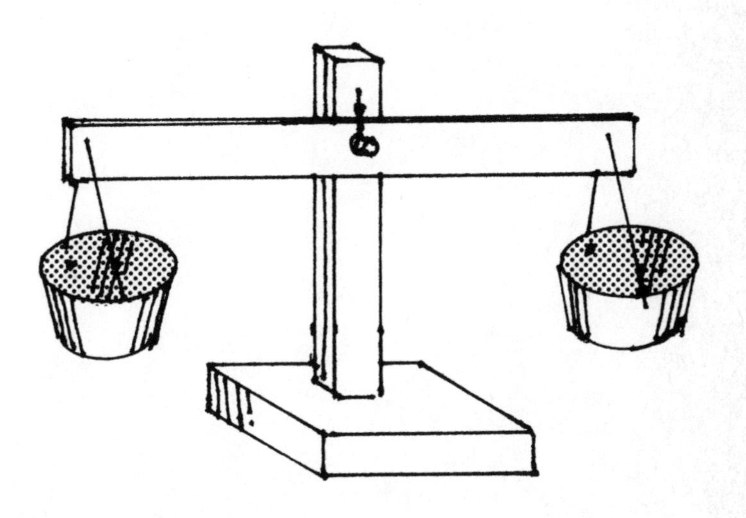

D

E

F

G

H

I

K

L

M

N

O

P

R

S

T

V

W